Medicinal Plants of Pilibhit Tiger Reserve (PTR) India

Authored by

Gopal Dixit

Department of Botany
Upadhi PG College
MJP Rohilkhand University
Pilibhit, India

Medicinal Plants of Pilibhit Tiger Reserve (PTR) India

Author: Gopal Dixit

ISBN (Online): 978-981-5313-95-6

ISBN (Print): 978-981-5313-96-3

ISBN (Paperback): 978-981-5313-97-0

First published in 2025.

need for a court order if at any point you breach any terms of this License Agreement. In no event will any delay or failure by Bentham Science Publishers in enforcing your compliance with this License Agreement constitute a waiver of any of its rights.

3. You acknowledge that you have read this License Agreement, and agree to be bound by its terms and conditions. To the extent that any other terms and conditions presented on any website of Bentham Science Publishers conflict with, or are inconsistent with, the terms and conditions set out in this License Agreement, you acknowledge that the terms and conditions set out in this License Agreement shall prevail.

Bentham Science Publishers Pte. Ltd.
80 Robinson Road #02-00
Singapore 068898
Singapore
Email: subscriptions@benthamscience.net

BENTHAM
SCIENCE

CONTENTS

PREFACE

The present manuscript in the form of a book is based on exhaustive field studies, surveys, questionnaires, and face-to-face interviews with certain ethical, rural people and herbal medical practitioners. These studies were carried out between 2015 to 2019 and covered the study area of the Indo-Nepal sub-Himalayan Terai International border region of Pilibhit Tiger Reserve (PTR), in Uttar Pradesh, India. In the present book, it has been discussed that 117 plants belonging to 44 families having medicinal value, are extensively used to treat more than 100 human and veterinary ailments . Few of these therapeutic practices are very new to the modern world, and most of these plants cure certain diseases in their daily life as these remedies are based on their generation's long traditional practices and passed on from one generation to another. Most of the therapeutic uses were successfully established after repeated trial and error methods by their ancestors.

In this way, the present book is a documentation and compilation of the literature based on their traditional knowledge about the therapeutic practices in their day-to-day life. The main motto of preparing this book is to enlighten the generations with old traditional knowledge about wild plants and their every possible use in treating most diseases. It has been kept in mind while proceeding with the studies on medicinal plants that the administration of many wild plants must have a positive impact on human beings and a negative correlation with most common ailments to draw an exact inference about certain useful plants and their products for their possible commercialization. Several wild and some cultivated plants were collected and thoroughly studied with the help of available literature. In most studies, repeated and replicated questioning methods have been taken to draw a conclusion about claims related to them.

Completion of this book would not have been possible without the unconditional and continuous blessings of **"THE ALMIGHTY"** and my parents. We are also thankful to forest officials of Pilibhit Tiger Reserve of district Pilibhit, UP, India for their cooperation with prior permission during the field visits. The help rendered by tribal heads, herbalists, traditional medical practitioners, and other rural and ethnic people in searching for wild plants and their possible usage as medicines has also been thankfully acknowledged.

The approach of experienced tribal and rural persons living around PTR was very positive toward scientific exploration and documentation of the natural plant products used to treat most of their daily ailments. The help of young tribal people was praiseworthy in collecting wild plants from natural sources in the vicinity of dense forests. Collected plants were identified with the help of available literature and under the guidelines set by our teacher and renowned Taxonomist, late Professor (Dr.) S.C. Sharma.

Last but not least, error finding in the spellings of the content of this book done by our sons Master Astitva and Tishye Dixit is of great importance in the completion of this task in time.

Gopal Dixit
Department of Botany
Upadhi PG College
MJP Rohilkhand University
Pilibhit, India

DEDICATION

This publication is

Dedicated to my beloved wife Late

Dr. Shilpa Vakshasya

St. Alloysius College , PILIBHIT, India

CHAPTER 1

Prologue

Abstract: The persistent and intricate symbiotic relationship between plants and human beings surpasses temporal and cultural borders, profoundly influencing the fundamental nature of human existence. Plants are crucial in providing nourishment, clothing, shelter, and therapeutic resources, making them essential components of everyday existence. The varied nature of this job transcends temporal and geographical boundaries, exhibiting relevance in various communities and historical epochs. The significant role of plants in preserving health and promoting health resonates with the principles of classical wisdom and traditional folk knowledge systems. As society progresses into the future, it is crucial to recognize and safeguard the invaluable knowledge of plants for the collective progress of humanity. The enduring significance of preserving our natural environment and the inherent wisdom it contains is demonstrated by the indisputable interconnectedness between plants and the existence of human beings. In a period characterized by significant technological advancements and profound societal changes, acknowledging and preserving our botanical history is crucial in promoting sustainable development. This chapter emphasizes the inherent connection between the Earth's ecological state and its population's welfare, emphasizing the imperative to save and diligently transmit this floral heritage to succeeding cohorts. The preservation of the deep interrelation between plants and human existence is evidence of the lasting wisdom inherent in the natural world, directing humans toward a future characterized by harmony and sustainability.

Keywords: Botanical wisdom, Interdependence, Symbiotic relationship, Sustainable progress.

INTRODUCTION

The relation between plants and human beings is as old as human civilization on the earth. Plants offer all three basic human needs: food, clothes, and shelter. Most of the things we use daily are gradual conversions of plant products. All herbivores' food comes from plants such as vegetables, fruits, cereals, pulses, *etc.* Even the large animal population depends on the producers, *i.e.*, green plants. In this way, we can say, "all flesh is grass".

The various sources of food, fibers, wood, drugs, beverages, oils, fats, waxes, *etc.*, are of plant origin. However, plants, in other ways, are responsible, indispensable, and unforgettable for the overall development of life on this earth.

Fibers also come from plants and are commonly used in weaving clothes, fabrics, ropes, threads, bags, and nets. Papers, cellulose, rayon, cellophane, and other critical industrial articles are made of fibers. Cotton, flax, jute, hemp, ramie, *etc.*, constitute important commercial fibers.

There exist several thousand medicinal plants all over the world. Most plants are known and utilized by herbal doctors and Ayurvedic *Vaidyas*. Many medicinal plants are found growing wild in various parts of the world. Only a few drug plants are cultivated. These plants are collected and prepared in crude in the indigenous way. The medicinal value of the plants producing drugs is due to some chemical substances present in the plant tissues, which can stimulate a definite physiological action in the human body. The most important chemical substances are alkaloids, carbon compounds, hydrogen, oxygen, nitrogen, glucosides, essential oils, fatty oils, resins, mucilage's, tannins, gums, *etc.* Some of these substances are also poisons and proteinaceous in nature.

Ethnic medicine is the best way to understand various aspects of Indian traditional knowledge systems. The most remarkable characteristic of the Indian medical tradition is that it prevails at two levels: the classical and folk systems. We refer to these systems as Ayurvedic, Siddha, and Unani under the classical system. They are characterized by institutionally trained practitioners, a body of text originating since ancient times, and highly developed theories to support the practices. These traditional medicinal systems encompass knowledge of life, health, and diseases of all living forms, including humans, animals, and plants. The branch dealing with traditional animal medicines is known as *Mrigayurveda* [1].

A rich textual base supports both Ayurveda and Unani systems of medical tradition. It is estimated that there are 10-30 million manuscripts in Sanskrit alone, many of them relating to medicine. Parallel to these systems, folk traditions exist, which has been transmitted orally in thousands of our villages throughout the ages. These folk traditions are rich and diverse, including specialized practitioners and home remedies for common ailments. These traditions include knowledge and beliefs regarding the relationship between food and health, yoga, and other preventive physical practices. Specialists work with specific diseases like fractured bone setting, snake-scorpion poisoning treatment, and birth attendance. A conservative estimate stipulates that around 70,000 traditional bone setters throughout the country attend to cover two-thirds of the fractures, as modern orthopedic facilities are few and mainly concentrated in cities and urban areas. Some 600,000 traditional birth attendants (dayees) perform home deliveries.

In India, using plants in medicine is as old as ancient Indian civilization. In Vedas and various other religious scriptures, their uses are frequently mentioned.

Ethnomedicinal knowledge about the uses of plants in different ailments and their cure among various ethnic groups in the Terai region of upper Gangetic plains are vast. This critical source of ethnomedicinal information is still unnoticed.

The ethnobotanical exploration of the remote forest areas of district Pilibhit has revealed that the locals use numerous plant species to treat various ailments. The tribal people have an immense knowledge about the medicinal use of plants. They can identify the plants as per their local names and their uses for treating different human and veterinary diseases.

Plants as Providers of Sustenance

Plants have always played an indisputable role in human history as the silent protectors of our food and survival. Forests are not merely the verdant backdrop of our surroundings but the essential source of our survival. This investigation delves into the deep importance of "Plants as Providers of Sustenance" as we explore how plants nourish and support us, encompassing nutritionally, culturally, and economically. Plants are the major source of essential nutrients our bodies need to grow and remain healthy. The nutritional benefits they provide are truly extraordinary. Fruits, vegetables, grains, legumes, and nuts are the fundamental components of a well-balanced diet. These natural resources are abundant in vital vitamins, minerals, and dietary fiber crucial for our health and overall health. Consider the diverse assortment of fruits, such as oranges, which are rich in vitamin C and enhance our immune system while revitalizing our skin. Spinach and kale, both leafy greens, offer a wide range of essential vitamins, such as A and K, which play crucial roles in maintaining good vision and promoting proper blood coagulation. Plant-based meals are rich in dietary fiber, which supports digestive health and aids in maintaining normal cholesterol levels. In addition to these vital nutrients, plants include a wealth of phytochemicals and antioxidants, which provide anti-inflammatory and anti-cancer benefits. The Mediterranean diet is well-known for its emphasis on plant-based foods such as olive oil, whole grains, and various fruits and vegetables. It is highly regarded for its ability to lower the risk of heart disease and other long-term illnesses.

The emergence of plant-based diets signifies a notable shift in human food consumption patterns. Plant-based diets have gained significance as individuals increasingly prioritize healthier and more sustainable nutritional choices. These diets prioritize the intake of plant nutrients while minimizing or excluding animal-based items. The reasons for this change are complex and include considerations of individual health, animal rights, and environmental impact. Plant-based diets encompass a range of dietary choices, from lacto-vegetarianism, which involves consuming dairy but abstaining from meat, to veganism, which completely avoids all animal products. The health benefits linked to certain dietary choices are extensively documented. They can minimize the likelihood of developing heart disease, lower blood pressure, and enhance the efficacy of weight management. A well-balanced plant-based diet promotes the consumption of vital nutrients while naturally maintaining low saturated fats and cholesterol levels.

Furthermore, the transition towards plant-based diets significantly impacts health and the ecosystem. The livestock business makes a substantial contribution to greenhouse gas emissions and deforestation. Therefore, opting for plant-based diets is an environmentally aware choice that effectively minimizes one's carbon footprint. These diets are also environmentally friendly regarding water usage, as they require less water for cultivation than meat production. The increasing acceptability of plant-based diets has led to a growing market for plant-based food products, ranging from highly realistic meat replacements to dairy-free alternatives. The rise of plant-based diets reflects a significant shift in our connection with food, highlighting the crucial connection between our dietary decisions and their effects on our health and the environment. This comprehensive introduction delves deeply into plants' pivotal role in nourishing people, encompassing nutritional aspects and the ever-changing dietary environment. The fundamental basis of human life is rooted in our reliance on plants as the principal means of nutrition [2]. Irrespective of our dietary inclinations, be they herbivorous, omnivorous, or carnivorous, it is universally acknowledged that plants are the fundamental basis of our food chain. Our daily dietary intake includes fruits, vegetables, cereals, legumes, and various plant-based foods [3, 4]. The organisms we consume as sustenance eventually depend on plants as their primary energy source and nourishment.

In summary, it can be concisely said that the phrase "all flesh is grass" emphasizes the deep and all-encompassing relationship between terrestrial existence and sustenance derived from plants. The increasing global population has led to a heightened reliance on plant agriculture to fulfill the needs of a rapidly expanding human populace. The progress made in agricultural processes, the diversification of crop kinds, and the expansion of cultivable territory have facilitated the ability of humanity to meet the growing food demands of an expanding population. Plants, ranging from core crops such as wheat and rice to a diverse array of fruits and vegetables, continue to serve as humanity's fundamental food source.

Nutritional Value of Plants

The nutritional importance of plants is indisputable. Fruits, vegetables, grains, and legumes are essential to a well-balanced diet. They are abundant reservoirs of essential vitamins, minerals, and dietary fiber, vital for optimal health. For example, fruits such as oranges and berries are rich in vitamin C, which enhances the immune system and supports skin health. Spinach and kale, leafy greens, have high levels of vitamins A and K. These vitamins are crucial for maintaining good vision and promoting proper blood coagulation. Moreover, plants serve as a commendable reservoir of dietary fiber, which facilitates the process of digestion and contributes to the maintenance of optimal cholesterol levels.

Moreover, the phytochemicals and antioxidants present in plants have a crucial impact on preventing chronic diseases. Flavonoids and polyphenols are chemicals that possess notable anti-inflammatory and anti-cancer characteristics. The Mediterranean diet, which strongly emphasizes plant-based foods such as olive oil, whole grains, and a diverse range of fruits and vegetables, is renowned for its capacity to lower the likelihood of developing heart disease and other long-term illnesses.

Plant-Based Diets

Plant-based diets have become increasingly popular as people strive for better and more environmentally friendly dietary options. These diets prioritize the intake of plant nutrients while minimizing or excluding animal products. The reasons for this change are diverse, frequently encompassing considerations regarding health, animal rights, and ecological viability.

Plant-based diets encompass a range of dietary practices, such as lacto-vegetarianism (which includes dairy but excludes meat) and veganism (which excludes all animal products). The health benefits linked to these diets are extensively documented. They can aid in mitigating the likelihood of developing heart disease, decreasing blood pressure, and efficiently managing weight. Furthermore, a diet primarily composed of plant-based foods can contain significant vital nutrients if properly proportioned. For instance, it guarantees sufficient consumption of fiber, vitamins, and antioxidants while naturally having a low content of saturated fats and cholesterol.

Plant-based diets have become popular not just because of their health benefits but also because of their favorable environmental impact. The livestock business plays a substantial role in the emission of greenhouse gases and the destruction of forests. By opting for plant-based alternatives, individuals can effectively diminish their carbon footprint. Plant-based diets can help conserve water resources, as they necessitate less water for cultivation than meat production.

With the increasing popularity of plant-based diets, there is a rising need for plant-based food products, ranging from meat substitutes to dairy-free alternatives. The emergence of plant-based diets signifies a notable transformation in our perception of food and its influence on our health and the environment.

Plants as a Source of Fibers and Materials

In addition to their role as a nutritional resource, plants contribute substantially to various aspects of human existence by providing fibers and materials [5]. Weaving, an ancient craft with a long history, involves converting plant fibers into various textile products such as garments, textiles, ropes, threads, bags, and nets. The industrial sector also heavily utilizes these adaptable fibers, employing them in manufacturing papers, cellulose, rayon, cellophane, and other crucial components. Cotton, flax, jute, hemp, ramie, and other plant-derived fibers represent significant classifications of commercially utilized fibers, thereby underscoring the exceptional adaptability of plants in addressing our wide-ranging requirements [6].

The Medicinal Bounty of Plants

Using plants for therapeutic reasons has been an indispensable component of human history and culture in every region of the world [7]. Herbalists, traditional healers, and practitioners of ancient medical systems like Ayurveda have used several thousand different species of therapeutic plants worldwide [8]. These medicinal plants can grow naturally in the wild, but others are purposefully grown in gardens because of their curative qualities. These plants can heal because their tissues contain chemical compounds [8]. These chemicals have unique physiological effects on the human body, so these plants have healing potential. These bioactive components cover a broad spectrum, including alkaloids, carbon compounds, hydrogen, oxygen, nitrogen, glucosides, essential oils, fatty oils, resins, mucilage's, tannins, gums, and many more substances [9]. It is essential to keep in mind that some of these compounds have the potential to be highly toxic, even though others have healing properties.

Traditional Medicine: A Multifaceted Heritage

Traditional medicine, firmly entrenched in medicinal plant application, offers profound insights into various areas of Indian traditional knowledge systems [10]. India's medicine field has a long and illustrious history incorporating traditional and folk practices. The classical systems, including Ayurveda, Siddha, and Unani, are characterized by practitioners who have undergone formal training, old literature, and well-developed theories that support their practices [8]. These systems not only focus on human health but also incorporate knowledge about the health of animals and plants in their analysis and recommendations. Mrigayurveda is the name given to the sub-division of Ayurveda that focuses on traditional animal medicine.

The textual heritage of Indian medicine is quite extensive [11]. Between 10 and 30 million manuscripts are believed to be written only in Sanskrit, most of which are related to medical topics [12]. These works have helped preserve information about life, health, and diseases, and they not only provide instruction for the care of humans, animals, and plants [13]. In addition to these traditional systems, folk traditions have flourished over time and have been passed down verbally *via* thousands of villages and hundreds of years. These folk traditions cover various activities and treatments for widespread diseases. They include medical professionals with particular training, such as traditional bone setters treating broken bones and traditional birth attendants (dayees) responsible for home births and deliveries [14]. Due to the lack of access to modern medical facilities in many outlying regions of India, these traditional medical practitioners are the primary source of medical care.

Plants in Indian Traditional Medicine: Ethnobotanical and Medicinal Significance

India, known for its varied cultures and customs, has a significant legacy in medicine. Ayurveda is a prominent holistic system of medicine in this subcontinent, distinguished by its extensive utilization of plants and herbs. Ayurveda, commonly known as the "science of life," is an enduring cultural legacy transmitted across generations and continues to impact modern medical methodologies significantly. Plants are central to this ancient medical system, utilizing their therapeutic powers to enhance health and facilitate healing. This investigation examines the deep connection between Indian culture, plants, and health in the context of "Plants in Indian Traditional Medicine." Ayurveda, an ancient medical system, has a history of more than 5,000 years, making it one of the oldest in the world. Ayurveda, derived from the ancient Indian writings called the Vedas, provides a holistic approach to health. Central to this concept is the conviction that maintaining equilibrium among the intellect, physique, and soul is crucial for holistic health.

Plants and herbs have a crucial role in Ayurveda. The system relies on thousands of plant varieties, each possessing distinct features and applications. The ancient Ayurvedic scriptures, such as the Charka Samhita and Sushrut Samhita, document an extensive range of medicinal herbs and their therapeutic use. These works include comprehensive explanations of the botanical attributes, preparation techniques, and medicinal properties of plants, serving as the foundation of Ayurvedic treatment. The core principle of Ayurveda revolves around the notion of the three doshas: Vata, Pitta, and Kapha. The doshas are manifestations of different combinations of the five elements—earth, water, fire, air, and ether—and are thought to regulate an individual's physical and mental makeup. The underlying cause of disease and illness is believed to be imbalances in the doshas. Ayurvedic practitioners utilize certain botanicals and medicinal plants to reinstate equilibrium in the doshas, fostering optimal health and overall health. As an illustration, the herb Ashwagandha *(Withania somnifera)* is employed in Ayurveda to alleviate Vatadosha. It is recognized for its adaptogenic characteristics, which aid the body in adjusting to stress and facilitating relaxation. Neem plants (*Azadirachta indica*) are employed to restore equilibrium to Pitta dosha and are recognized for their antibacterial and anti-inflammatory characteristics. Ginger (*Zingiber officinale*) and Turmeric (*Curcuma longa*) are commonly used for Kaphadosha due to their digestive and metabolic stimulating properties. The therapeutic efficacy of plants in Ayurveda transcends the mere equilibrium of doshas. Ayurvedic practitioners frequently recommend specialized herbal mixtures, "rasayanas," to revitalize the body and improve longevity. These formulations consist of a combination of meticulously chosen herbs designed to target various health issues. Chyawanprash, an acclaimed Ayurvedic rasayana, is composed of a blend of herbs, such as Amla (*Emblica officinalis*), Ashwagandha, and Guduchi (*Tinospora cordifolia*), among others. It is famous for its ability to enhance the immune system and is commonly used to improve overall health and energy.

Ayurveda extends beyond merely using herbs exclusively for therapeutic intentions; it encompasses one's way of life. The method suggests daily schedules, food selections, and herbal treatments that promote overall health and wellness. Engaging in the daily routine of "oil pulling" with sesame oil to maintain oral health, incorporating ghee (clarified butter) into one's diet for its nourishing qualities, and utilizing spices such as cumin and coriander to aid digestion are all integral components of the Ayurvedic way of life. Ayurveda emphasizes the significance of mindfulness, meditation, and yoga. These practices are essential for preserving harmony in mind, body, and spirit. There are worries over the excessive harvesting and loss of habitats due to the extensive utilization of therapeutic plants in Ayurveda. Many of these plant species are currently facing the threat of extinction. Preserving the biodiversity of India's plant resources requires implementing conservation measures and sustainable harvesting practices.

In addition, although Ayurveda has gained global acclaim and is highly regarded for its comprehensive methodology, there are ongoing difficulties in establishing uniformity and oversight for Ayurvedic products. The ongoing discussion revolves around the quality control and safety of herbal treatments. Plants have been fundamental to Indian culture and traditional healing for thousands of years. Ayurveda, an ancient discipline, represents the deep interrelation between humans and the botanical realm. The holistic approach to health, which is based on the utilization of medicinal plants, continues to have a significant impact on modern medicine and provides a timeless heritage that serves as a reminder of the therapeutic potential of nature. As we contemplate the future, preserving and safeguarding these botanical resources and guaranteeing the longevity of Ayurvedic knowledge for future generations is imperative. The incorporation of plants in Indian traditional medicine exemplifies the lasting importance of nature in human health, and it catalyzes the amalgamation of traditional and modern medical approaches in the quest for a healthier global society.

The use of plants in traditional Indian medicine has a long and illustrious history in the country, dating back to when ancient writings like the Vedas were written [15]. There is a vast body of information regarding the use of plants in ethnomedicine to treat various illnesses among numerous ethnic groups in the Terai region of the upper Gangetic plains. Specifically, this body of knowledge pertains to the treatment of various ailments. Even though it is an excellent source of ethnomedicinal information, relatively few people are aware of it, yet it has enormous potential applications in contemporary healthcare [16, 17].

Overview of the Indo-Nepal Sub-Himalayan International Border Region of PTR

The Indo-Nepal sub-Himalayan border region, where Pilibhit Tiger Reserve is located, forms a part of Terai Arc landscape. This lowland belt, nestled at the foothills of the Himalayas, stretches across northern India and southern Nepal the region is characterized by its unique ecological and geographical features, including dense forests, grasslands, and wetlands that support a rich biodiversity.

Pilibhit Tiger Reserve is situated in the western part of Uttar Pradesh, India, along the international border with Nepal. The reserve is known for its extensive Sal forests diverse wildlife, and vibrant ecosystems. The area is home to several endangered species, such as the Royal Bengal tiger, Indian Leopard, and swamp deer, as well as a wide variety of bird species.

The border region serves as a vital ecological corridor for wildlife, promoting connectivity between protected areas in both India and Nepal. This cross-border landscape also holds significant cultural and ethnobotanical importance, with local communities on both sides relying on the forest resources for traditional medicinal practices agriculture, and sustenance. The region's flora includes a vast array of medicinal plants, many of which are integral to indigenous healthcare systems.

The blend of rich biodiversity, traditional knowledge, and international collaboration makes the Indo-Nepal border region of Pilibhit Tiger Reserve a critical area for both conservation and ethnobotanical research.

Ethnobotanical Discoveries in District Pilibhit: A window into Traditional Wisdom

Research on the inaccessible forested regions of district Pilibhit has unearthed a veritable treasure trove of traditional knowledge about using plants for therapeutic reasons [18]. Communities on the ground, especially notably indigenous peoples and communities, have an in-depth knowledge of the therapeutic potential of various plants. They can recognize plants by their regional names and know how these plants can be used to cure various illnesses. These tribes' healthcare methods continue to use the indigenous knowledge handed down through the generations significantly. The district Pilibhit nestled along the Indo-Nepal border and hoe to Pilibhit Tiger Reserve, PTR, is rich in ethnobotanical knowledge passed down through generations. Local communities, including indigenous tribes like Tharus, Van Gujjars and Kanjars, and rural populations, have long relied on the region's diverse flora for medicinal, nutritional, and cultural purposes.

Ethnobotanical discoveries here reveal the extensive use of plants for treating ailments such as digestive troubles, respiratory ailments, skin disorders, and many more. These findings highlight the importance of traditional wisdom in the sustainable use of medicinal plants and underscore the urgent need to preserve this knowledge in the face of modern development and ecological challenges.

Scope and Purpose of the Study

The complex and persistent symbiotic connection between plants and human beings surpasses temporal and cultural boundaries. Plants play a multifaceted role in human existence, offering sustenance, apparel, habitation, and medicinal resources that significantly impact several aspects of our daily lives. The importance of plants in preserving and promoting health is emphasized in classical and traditional folk knowledge systems. As society progresses into the future, it is crucial to recognize and safeguard this vital botanical knowledge for the advancement of humanity. The ongoing significance of preserving our natural environment and the information it imparts is exemplified by the interdependence between plants and human life. The scope of this book is to document and explore the diverse medicinal plant species growing in the Indo-Nepal border region of Pilibhit Tiger Reserve, PTR. It aims to highlight the traditional knowledge of local communities, the therapeutic uses of these plants, and their ecological significance. The purpose of bringing this manuscript is to provide a comprehensive resource for researchers, conservationists, and practitioners, while promoting the conservation and sustainable use of medicinal plant biodiversity in this unique cross border region of the province.

CHAPTER 2

Study Area Details

Abstract: District Pilibhit, situated in the foothill region of the Indo-Nepal Terai, is recognized as the most affluent biodiversity hotspot within the state of Uttar Pradesh. The municipality, despite its recent establishment, encompasses a historic settlement known as 'Old Pilibhit' situated alongside the Khakra River, which holds significant cultural and historical value due to its association with the Banjaras of the Periya clan. The etymology of the term 'Pilibhit' implies a potential association with 'PeriyaBhit,' which denotes the village mound of the Periyas. Alternatively, it could be derived from a historical reference to a yellow mud wall that used to encompass the district. Pilibhit is geographically located in the Shivalik foothills of the Himalayas and benefits from its accessibility *via* road networks and a meter-gauge railway line, facilitating connections with adjacent districts in the states of Uttar Pradesh and Uttarakhand. The district's importance is further elevated by a multitude of tourism and historical attractions. The Pilibhit Tiger Reserve (PTR) is worthy of attention since it was officially designated by the Central government in 2014. It encompasses an area of 730.24 square kilometers located within the densely forested region of Pilibhit. The study examines the geographic location, topographic features, climatic patterns, and vegetation composition of the region. The primary objective of this extensive investigation is to reveal the complex fabric of Pilibhit, highlighting its ecological abundance, historical origins, and its significant function as a wildlife sanctuary, particularly as the third tiger reserve in the state of Uttar Pradesh.

Keywords: Biodiversity hotspot, Historical significance, Pilibhit tiger reserve, Shivalik foothills, Terai region.

WORKSPACE

District Pilibhit, situated in the foothill area of the Indo Nepal Terai international border region and the richest hotspot of biodiversity in UP State, has been selected for the present study.

The present town is of comparatively recent origin, but there is still a village known as 'Old Pilibhit' standing on the left bank of the Khakra river about 5 Km. to the north-east near the road to Neoria. The Banjaras of the Periya clan had always occupied this village. It is supposed that Pilibhit is the corruption of Periya

Bhit or the village mound of the Periyas, and also that the name Pilibhit has been derived from a yellow mud wall that once surrounded the district.

Pilibhit is situated in the Shivalik foothills of the Himalayas, well connected by road and broad gauge railway line of the North Eastern Railways to its adjoining Districts, *viz.*, Bareilly, Shahjahanpur and Lakhimpur Khiri of Uttar Pradesh and Udham Singh Nagar of Uttarakhand. Many tourist and Historical places increase the overall importance of the district.

Pilibhit Tiger Reserve (PTR) was declared by the central government in 2014, and is the third tiger reserve in the state of Uttar Pradesh, sharing open international boundary with neighboring country Nepal.

From the study point of view, the Study Area has been described under the following subheads:

• Geographical Situation
• Topography
• Climatic Conditions
• Vegetation

GEOGRAPHICAL SITUATION

The district of Pilibhit is the north-eastern most district of Rohilkhand division. It is situated in the foothill region of the terai of upper Gangetic plains, lying between the parallels 28^0 54' - 28^0 60' N latitude and 79^0 37'-88^0 27' E longitude at an elevation of 183.870 m above MSL (Fig. **1**). It is situated in the foothills of Shivaliks of the Himalayas and consists of various mountain ranges. On the north is the district Udham Singh Nagar and the territory of Nepal; on the south lies the Shahjahanpur district; on the east, the district is flanked for a short distance by district Kheri and the remaining distance by the Shahjahanpur district and on the west the district of Bareilly.

The north eastern part of Pilibhit is mostly constituted by the Terai area enriched with several rivers and lakes.

RIVERS & WATER SOURCES

• Sharda
• Chauka or Chuka
• Gomti
• Khannaut
• Mala
• Deoha

- Lohia
- Khakra
- Rapatua
- Sundaria & Kailas
- Apsara
- Pangaili
- Fulhar Lake (Madhotanda)
- Mahadev lake (Jamunia)
- Bari lake (Bithora Kalan)
- Anwar ganj lake (Anwarganj)
- Balpur lake (Balpurmandalia)

Fig. (1). The geographical area of Pilibhit is 3765.7 sq. Km which is 1.3% of the total geographical area of Uttar Pradesh. Of the district's total area, 99.1% is rural, and 0.9% is urban. In terms of the area, it is the smallest compared to other districts of Bareilly Block.

River Gomti originates from Fulhar lake of Madhotanda; River Khannaut from Madhav lake; River Khera from Anwarganj Lake, Apsara from Badhpur Lake. Besides these, River Mala, Deoha, Katna, Khakra, Kailash, Pangahli originate from the mountains and flow from North to South, whereas Sharda and Chuka flow in the Northeastern part.

Pilibhit has a dense forest cover on about 21.97% (78478 hectares) of the total area falling in the Puranpur and Pilibhit Tehsils. Two distinct forest divisions exist in the district *i.e.* Pilibhit forest division and Social forestry division, Pilibhit. The former is engaged in proper care and maintenance of the reserved forest areas of the district. In contrast, the latter looks after the 941.62 ha reserved forest area in Puranpur Tehsil along with the forestation along the roadsides, railways and Panchayat lands. Five Forest ranges fall under the Pilibhit forest division, *viz.*, Mala, Mahof, Haripur, Barahi and Deoria, whereas Social Forestry division covers three ranges-Pilibhit, Bisalpur and Puranpur.

Timber yielding trees like Sal (*Shorea robusta*), Shisham (*Dalbergia sissoo*), and mango (*Mangifera indica*), katha (*Acacia catechu*), herbs, and honey are found in these forests. In the terai area there is abundance of grass like *baib* and *munj*. Many wild animals are found in the Mala, Sharda and Chuka forests.

Wheat, rice, maize, jowar, bajra, gram, sesame and groundnut are the major crops found in the plains, with sugarcane as the basic cash crop. In the grass abundant areas, animal husbandary is the prime business. Sugarmills have been set up in Majhola, Bisalpur, Puranpur and Pilibhit. Cottage industries include mat weaving, wood industries and collection of medicinal plants from the forests. Cereals, Jaggary, sugar are sold in the mandies of Pilibhit, Bisalpur, Majhola, Neuria and Madhotanda, which are well connected by roads.

To facilitate transportation in the district there are state highways (137.845 Km); prime district Roads (92.38 Km); other district roads (329.22 Km); and village roads (729.86 Km).

- There are three Tehsils in the district- Pilibhit, Puranpur and Bisalpur, with seven development Blocks- Marauri, Lalaurikhera, Amaria, Barkhera, Bilsanda, Puranpur and Bisalpur.
- Pilibhit, Puranpur and Bisalpur are three Municipal Counsils with six Municipal Panchayats-Neuria, Guladia, Bhindara, Majhola, Jahanabad and Barkhera.
- Out of 1443 total villages 1293 are populated, while the number of Forest Villages is 9.
- There are six rural and eight urban Police Stations in the District.
- The broad gauge railway line is about 132 Km long with 20 railway stations (with halts).
- Pilibhit has 62 nationalized banks, 19 rural branches, 15 Cooperative Banks, 5 Cooperative farms and village Development Branches.
- Thousands of biogas plants are still in operation till today for the sustainable eco

friendly household cooking.

TOPOGRAPHY

In its general appearance, the district presents diverse features and may be divided into several distinct tracts topographically. In the north and north-west, the tract continues the *Terai*. The southern portion of the Bisalpur Tehsil is similar in most respect to the adjacent tract of Bareilly and Shahjahanpur. The eastern and smaller section approximates instead undeveloped forest.

Areas of Khiri, though with the spread of cultivation, the dissimilarity between Puranpur and the rest of the district is gradually becoming less marked.

All the district's rivers flow from north to south, except Sharda, which flows from North-West to South East, indicating that the north eastern part of the district is sloppy towards South West. The rest of the District is sloppy towards North-South. River Gomti originates from Madhotanda of Puranpur Tehsil as a small river and attains a large form when reaches the State capital Lucknow. With heavy rainfall on steep slope, gravel, sand and clay is eroded by water (or by wind) from rock surface, are generally brought down by rivers and streams.

Most of the area of the district consists of the soil eroded by various rivers. This soil varies from sandy to clayey. At many places the soil is sandy & usar. In the district's Puranpur area, the gravel soil is mostly eroded by River Sharda from the adjoining mountains. The eastern and central part is not plain, central and northern of which is sandy. River Sharda is responsible for soil erosion in the district, the flow of which increases due to heavy rains or water released by ShardaSagar Dam. The foothill area is sloppy and is a major victim of soil erosion. The villages situated at the bank of River Sharda are always threatened by massive floods & some of them have even lost their identity. The local departments, authorities and villagers cannot control the river's flow and it even changes its route.

The major portion of the district consists of soil eroded by rivers. This plain is ordinarily fertile. Paddy, sugarcane, wheat, gram, til, bajra, urad & moong are cultivated on a large scale along with fruits and vegetables like mango, jamun, pineapple, banana, litchi, tomato, pumpkin, cucumber, *etc.*

Water Channels (Canals)

Other than natural rivers, various canals are also helpful for irrigation purposes. Of the district's canal system, the total length of the Sharda canal system & Rohilkhand canal system is 1002 Km. Sharda canal was built from Banbasa of

District Udhamsingh Nagar. The place from which Khiri.

and Hardoi branches are released is Bifurcation, a tourist place. The Sharda canal is the main canal of the district, the others being its branches. The names and lengths of various canal branches are given below in Table **1**.

Table 1. Canal network and length in kilometers: contributions to irrigation infrastructure (1928-1957).

Name of Channel (With Year of Functioning)	Length in KM.
Sharda canal (1928)	12.64
Hardoi branch (1928)	36.80
Kheri branch (1928)	31.20
ShardaSagar feeder (1957)	03.90
Outlet channel (1957)	03.23
Subsidiary Hardoi branch (1957)	21.55

CLIMATIC CONDITIONS

Meteorological influences chiefly determine climate of any region as relative humidity of air, temperature, wind pressure and evaporation rates. Climatic factors characterize, in general a particular region. These factors are concerned mainly with the aerial environment, including light, air temperature, humidity, rainfall, gaseous atmosphere component, and wind. The climate of a particular region is greatly affected by the topographic factors and forms a local, even a microclimate of a region.

So is the climate of Pilibhit, greatly influenced by the area's topography. As Pilibhit is situated in the foothills of the Shivaliks of the Himalayas, its climate varies greatly. The climate is characterized by three distinct seasons, *i.e.* summer, winter, and rainy. The average rainfall is about 1256 mm, whereas the normal is 738 mm.

The maximum and minimum temperature is 45.5 ^{0}C & 3.6 ^{0}C, respectively in June and January with a relative humidity of 76.2%.

The terai is greatly influenced by the climate of the hilly region of Uttarakhand, as when Uttarakhand receives solid rains, hails or even snow, the cold waves from the hills greatly affect the terai. The cold waves come down towards the plains of Pilibhit and create its microclimate.

It has often been observed that when there is no possibility of rain forecasted, the weather unexpectedly becomes cloudy and it starts raining, whether for a few minutes or even for some hours, thus changing the weather in a couple of minutes.

Summers commence from March end and last till August followed by rainy season. November, December, and January are marked with extreme cold conditions and fogs that seldom last for days or even weeks. Mostly, there remains a dense fog from morning to noon that begins late in the evening.

In the winters, the condition becomes colder when the area receives cold waves from the adjoining hilly area.

VEGETATION

The forests of Pilibhit district belong to tropical moist deciduous type. These are mainly found in Haripur, Mala, and Varahi forest ranges. Trees, shrubs and perennial herbs represent the permanent vegetation of the forests. The herbaceous flora is mainly seasonal and appears during the rainy season. Based on the floristic composition, the vegetation of the district may be divided into following types:

• Mixed forests
• Swamp forests
• Grasslands

Mixed Forests

This type occupies large stretches of forest division and occurs on higher and more stable alluvial terraces. The main species are *Adina cordifolia* (Roxb.), Hook f. ex Brandis; *Albizzia procera* (Roxb.) Benth; *Ficus semicordata* Buch-Ham.; *F. racemosa* Linn.; *Tectonagrandis* Linn. f.; *Dalbergia sissoo* Roxb.; *Bombax ceiba* Linn; *Acacia catechu* Willd.; *Mitragyna parviflora* Korth; *Emblica officinalis* Gaertn.; *Gmelina arborea* Roxb.; *Litsea glutinosa* (Lous) C.B. Robins; *Litsea monopetala* (Roxb.) Pers; *Bauhinia racemosa* Lamk; *Bauhinia variegate* Linn.; *Sterculia villosa* (Roxb.); *Ziziphus mauritiana* lamk.; *Butea monosperma* Lamk., *etc.*

The common shrubs are *Carrisa opaca* Stapf. Ex. Haines; *Leea* sps; *Calotropis gigantea* (Linn.) R. Br.; *Calotropis procera* (Ait) R. Br., *etc.* Common climbers include *Ampellosiccus* latifolia (Roxb.) Planch; *Tinospora cordifolia* (Willd.) Miers ex Hook f. Thomas; *Ventilago denticulate* Willd, *etc.*

Common grasses are *Saccharum spontaneum* Linn.; *Apluda mutica* Linn.; *Heteropogon contortus* (Linn.) Beauv.; *Stellaria glauca* (Linn.) Beauv., *etc.*

Swamp Forests

The vegetation occurring in low-lying water-logged area along nallas, river beds and stream beds which remain under water during rain and some times many months thereafter develops into a swamp forest. The aeration of soil is usually poor. The soil is rich in humus and consists of fine clay it belongs to either Barringtonia swamp forest or Syzygium swamp forests. The former type is seen along the swampier areas while the later comparatively dry areas. At certain places these forests get mixed together.

Grasslands

The grassland occupies large stretches and lie scattered in between forests. The common grasses are *Themeda arundinacea* (Roxb.) Ridley; *Apluda muricata* Linn.; *Vetiveria zizanioides* (Linn.) Nash.; *Saccharum benghalense* Retz; *S. spontaneum* Linn.; *Imperata cylinderica* (Linn.) Beauv; *Eulaliopsis binata* (Retz.) Hubb., *etc.*

Seasonal Vegetation

The seasonal vegetation comprises herbs and under shrubs growing on various habitats during the rainy winter and summer.

Inference

In conclusion, district Pilibhit is a captivating blend of natural richness and historical significance. Its status as the richest biodiversity hotspot in Uttar Pradesh, coupled with establishing the Pilibhit Tiger Reserve in 2014, underscores the region's commitment to environmental preservation. The town's historical roots, reflected in the existence of 'Old Pilibhit,' add a layer of cultural depth, suggesting a connection to the Banjaras of the Periya clan. The etymological insights into 'Pilibhit' further contribute to the district's mystique, possibly linked to ancient village mounds or protective mud walls.

Geographically situated in the Indo-Nepal Terai region, Pilibhit's proximity to the Shivalik foothills of the Himalayas enhances its allure. The district's connectivity through roads and a meter-gauge railway line facilitates transportation and underscores its accessibility to neighboring districts in Uttar Pradesh and Uttarakhand. Including various tourist and historical sites elevates Pilibhit's overall importance in the region. As the study delves into the geographical situation, topography, climatic conditions, and vegetation, it unravels the layers of complexity that contribute to Pilibhit's unique identity. The district stands as a testament to the delicate balance between conservation and development, symbolized by the establishing the Pilibhit Tiger Reserve. This holistic understanding of Pilibhit serves as a valuable foundation for ongoing environmental protection and sustainable management efforts, ensuring that the district continues to thrive as a harmonious intersection of nature and culture.

CHAPTER 3

Enumeration of Families of Medicinal Plants

Abstract: This study investigates and enlists several plant families in Pilibhit, a foothill district of the Indo-Nepal sub-Himalayan International border region of Terai and one of Uttar Pradesh's richest biodiversity hotspots is here. The study will cover these plant groups' botanical traits, historical usage, and ecological importance. Pilibhit district's unique location and different ecosystems make it excellent for plant biodiversity research. It has been investigated that many plant families, including therapeutic herbs and economically important species. This study details the distribution of several regional plant groups, their ecological roles, and their cultural significance. The research examines medicinal, ethnobotanical, and ethnoveterinary uses of these plant families and their effects on local, healthcare, and traditional practices. It also examines their ecological services, including habitat, food, and soil stabilization, and how they support local biodiversity and ecosystem functioning. This study highlights the Pilibhit district's botanical diversity and the importance of these plant families for conservation, sustainable resource management, and traditional knowledge preservation. Understanding the complex relationships between humans and plants in this biodiversity hotspot is essential for biodiversity conservation and community health.

Keywords: Biodiversity, Healthcare, Medicinal plants, Plant families, Traditional knowledge.

FAMILY ACANTHACEAE

Introduction

The Acanthaceae, or Acanthus, family of flowering plants is incredibly varied and essential in horticulture and traditional medicine [19]. Approximately 250 genera and nearly 2,500 species give this plant family its reputation for diversity [20]. Plants in the family Acanthaceae are most common in the tropics and subtropics, and they thrive well in a broad variety of environments. One of the most defining characteristics of Acanthaceae plants is their bi-labiate blooms, which have two lips that vary in shape and color [21]. The allure of their flowers has made some members of the Acanthaceae family standard fixtures in landscaped decorative areas. Not only are many members of the Acanthaceae family beautiful to look at, but they have also been used for centuries in the traditional medicine systems of many different cultures, including those of Asia, Africa, and the Americas [22].

The healing powers of these plants have been used to cure a wide variety of illnesses, from fevers to gastrointestinal issues to skin problems. Notable examples are the Ayurvedic herb *Andrographis paniculata* and the traditional Chinese medicine staple *Justicia adhatoda*. Some members of the Acanthaceae family have also attracted the interest of scientists due to the possible medicinal chemicals they contain [23]. Their secondary metabolites, including alkaloids and flavonoids, are still being researched for their potential therapeutic uses. In conclusion, the Acanthaceae family is an exciting collection of plants because of their beauty and medical value. Their ecological and pharmacological potential is underappreciated, but their historical significance in traditional medicine systems makes them a crucial subject of study and conservation efforts.

The plants belonging to the Acanthaceae family have been acknowledged for their possible therapeutic advantages in several traditional medicinal practices. Although they are significant in traditional medicines, seeking advice from healthcare professionals or traditional healers knowledgeable about their usage is recommended to guarantee a secure and efficient treatment. They continue scientific research endeavors to authenticate and enhance comprehension of the precise chemicals accountable for their therapeutic impacts.

• *Justicia gendarussa* **Burm. f.**

Local Name: Adusa

Botanical Characteristics: Shrubs of 1-2.5 m height with opposite ascending branches and 10-20 lance-shaped leaves that are simple, opposite7-19 cm long and 4-7 cm wide.

Distribution: Wild

Therapeutic Usage:

Ethnopharmacology:

• Traditionally, it treats many health ailments such as respiratory conditions, fever, and digestive disorders. The leaves of Malabar Nut are reputed to possess antibacterial and anti-inflammatory attributes, making them valuable in treating respiratory infections and gastrointestinal ailments.
• The plant's extracts are integrated into herbal compositions in certain traditional medical treatments.

• *Ruellia tuberosa* **L.**

Local Name: Bhukanda

Botanical Characteristics: A perennial herb of 45cm tall, the stem is erect leaves are oblong, and hairy, flowers are pale blue, fruits are dry capsules, and roots are thick, cylindrical, and finger-like.

Distribution: Wild

Therapeutic Usage:

Ethnopharmacology:

• The plant has been traditionally used for medicinal purposes. The tubers of this plant are utilized in traditional medicine in certain cultures due to their probable diuretic qualities.
• Its root is believed to enhance the process of diuresis and can be employed to relieve problems associated with edema.

• *Barleria prionitis* L.

Local Name: Vajradanti

Botanical Characteristics: Shrub, leaves are elliptical, 5-20 mm long, with spines at the axis, flowers are yellow, sessile, capsule is two-celled, ovate-lanceolate, and 10-20 cm long with a sharp pointed beak about 6 mm long.

Distribution: Wild

Therapeutic Usage:

Ethnopharmacology:

• It is utilized in traditional medicinal practices in certain areas. The plant has historically been used externally on wounds and skin disorders due to its potential as an antibacterial and anti-inflammatory agent.

• *Adhatoda zeylanica* Medic. Syn. *Adhatoda vasica* Nees in Wall

Local Name: Vasaka

Botanical Characteristics: Shrubs, leaves opposite, simple, elliptic, lanceolate; flowers in thyristor spike, white, bilobed; fruits capsules; seeds tubercular varicose.

Distribution: Commonly grown as a hedge plant.

Therapeutic Usage:

Ethnopharmacology:

- Syrup made from Fresh and green flowering twigs is advised for cough and cold.
- Leaf powder is boiled in sesame oil and applied to the affected part of skin, till cure.
- To treat tuberculosis, a decoction of leaves along with jaggary and water is kept in an earthen pot for a month and is taken orally in sufficient quantity.
- Decoction of stem bark is used three times a day for expelling intestinal worms in children.

Ethnoveterinary:

Tribals give root bark decoction with black pepper powder in 5:2 to the cattle for safe discharge of the fetus after the delivery.

- *Peristrophe bicalyculata* (Retz.) Nees

Local Name: Hadjor

Botanical Characteristics: Erect, much branched, hispid herbs or under shrubs with angled branches; leaves opposite, petiolate, ovate, acute or acuminate, sparsely hairy; flowers pink; corolla blipped stamens two; fruit capsules oblong, pointed.

Distribution: Occurs commonly in shady places in open areas and as an undergrowth of an orchard.

Therapeutic Usage:

Ethnopharmacology:

- Paste made from green plants is bandaged on fractured bone for 20 days.
- Root paste is applied over the eczema-affected parts of the body.
- One teaspoon of decoction of green leaves is given twice a day to the patients of insects and snakebites.
- In rheumatism, root paste is applied externally twice daily for 15 days.

FAMILY AMARANTHACEAE

Introduction

The Amaranthaceae family, or amaranth family, comprises around 175 genera and roughly 2,500 species of flowering plants [24]. These plants can be found in various habitats, from arid regions to grasslands and tropical rainforests, demonstrating their remarkable flexibility. Ethnogynaecology, Ethnopharmacology, and ethnoveterinary medicine have much to learn from this family's extensive archive of traditional knowledge [25]. Species of the family Amaranthaceae have long been valued by indigenous communities for their medicinal and other unique qualities. Certain varieties of amaranth have been used in Ethnogynaecology to treat various conditions affecting women's health and reproduction, including menstrual regulation, increased fertility, and complications during pregnancy [26]. Amaranthaceae species have bioactive substances used in traditional medicine to treat gastrointestinal illnesses, skin diseases, and parasitic infections; these plants are the focus of Ethnopharmacology [27]. In rural and indigenous cultures, where there is a direct connection between humans and animals, these herbs are used in ethnoveterinary medicine to cure cattle illnesses. The importance of maintaining traditional knowledge of these plants is highlighted by the wide variety of uses to which members of the Amaranthaceae family have been put throughout history [28].

• *Amaranthus viridis* **L.**

Local Name: Chaulai

Botanical Characteristics: Herb, stem is erect, green, and branched; leaves are ovate, 20-80 mm long, petioles can be 30-80 mm long. Flowers are green and unisexual, fruits are green and wrinkled, and seeds are brown and black.

Distribution: Wild

Therapeutic Usage:

Ethnopharmacology:

- It is a highly nourishing leafy green vegetable that has long been a fundamental ingredient in various traditional cuisines across the globe. It is widely recognized for its abundant supply of vitamins, minerals, and dietary fiber. This highly nutritious food is abundant in vital elements such as vitamin A, vitamin C, calcium, and iron. It has been highly valued for its ability to enhance general health and vigor. The leaves are commonly used in soups, stews, and salads in traditional culinary techniques. They offer a delectable and highly nutritious enhancement to meals. The plant's adaptability and outstanding nutritional composition render it a favored option for individuals seeking to enhance their dietary consumption of essential vitamins and minerals.
- It has a historical background in traditional medicine, being utilized for its potential therapeutic advantages. Nevertheless, it is crucial to acknowledge that the main significance of Green Amaranth resides in its nutritional benefits.

Whether ingested in soups, sautéed, or as a salad component, it provides a nutritious and delicious means of fulfilling necessary dietary needs.

• *Alternanthera sessilis* (L.) R.br. ex. DC

Local Name: Gurundi

Botanical Characteristics: Perennial herb up to 10-45 cm tall, elongated leaves, flowers small white, stalkless, fruits egg-shaped.

Therapeutic Usage:

Ethnopharmacology:

• It is a very adaptable plant traditionally utilized in several areas for its culinary and therapeutic properties. It is reputed to offer anti-inflammatory and antioxidant qualities, rendering it a significant botanical resource in traditional medicinal practices. It has been employed to mitigate fever, regulate skin problems, and address gastrointestinal pain. The soft leaves are commonly used in culinary preparations, such as stir-fries and soups. These recipes are enhanced by the addition of a refreshing and subtly acidic taste, as well as the inclusion of vital nutrients.

• It possesses antioxidant qualities, rendering it a desirable dietary inclusion that may assist in mitigating oxidative stress within the body. Furthermore, the reason behind its conventional application for digestive issues can be ascribed to its high dietary fiber content, which helps promote digestive health. Although it is commonly used in traditional treatments and culinary practices, it is crucial to get advice from healthcare professionals or traditional practitioners to ensure its safe and effective use.

• *Gomphrena globosa* L.

Local Name: Gol Manika

Botanical Characteristics: Annual, 12-24 inch tall, leaves long and slender, opposite, white, wooly, flowers tiny, white, fruits small, bloon-like, stem erect and branched.

Distribution: Cultivated

Therapeutic Usage:

Ethnopharmacology:

- It is a unique plant highly regarded for its decorative attributes. Nevertheless, it also retains conventional use derived from its diuretic effects. Globe Amaranth is thought to possess diuretic properties, which can stimulate urination and potentially be employed in treating illnesses associated with fluid retention. In several ancient medical systems, it has been utilized to treat edema and other illnesses marked by an excessive accumulation of fluids in the body.
- It is visually captivating, showcasing spherical clusters of vibrant bracts that have rendered it a favored option for floral compositions and decorative gardens. The plant's vibrant and enduring bracts are available in various hues like purple, red, and white. Although Globe Amaranth has traditional applications in herbal medicine, its greatest value lies in its decorative attractiveness. Consulting healthcare professionals is crucial when contemplating the usage of this substance for therapeutic reasons, as it guarantees both safety and efficacy in treating fluid-related ailments.

• *Achyranthes aspera* **var.** *aspera* L.

Local Name: Chirchita

Botanical Characteristics: Erect, pubescent, annual herbs; leaves are opposite, simple, ovate, or elliptic; flowers in terminal spikes, greenish-white, pink, deflexed; fruits are utricles; and seeds are oblong, brown.

Distribution: Very common weed in waste places in association with common herbs and undershrubs.

Therapeutic Usage:

Ethnogynaecology:

The root decoction is given orally to women after menstrual cycle as a contraceptive.

Ethnopharmacology:

- External application of root decoction is effective against insects, scorpions and snake bites.
- 1 teaspoon of seed powder and black pepper is given thrice a day for 15 days for rheumatism, gout, and arthritis. Leaf paste with mustard oil and camphor is applied locally for rheumatism and gout.

- Tribal people advised the plant to paste with the powder of *Piper longum in* 3:2 ratios, twice a day for 7 days, to the patients of a mad dog bite. They do not allow sour products during the treatment.
- Powdered root bark with sugar is given twice daily to cure diarrhea and dysentery.
- The treatment of spermatorrhea root paste with black pepper is given orally for 15 days.

Ethnoveterinary:

Tribal and rural people keep 7 roots on horns for easy and safe delivery of buffaloes.

- *Achyranthes aspera* var. *porphyristachia* (Linn.) Hook f.

Local Name: Chirchiri

Botanical Characteristics: Erect or sub-scandent herbs or under shrubs; leaves ovate, elliptic, membranous, acute or acuminate, glabrous; flowers greenish white, deflexed, in terminal slender spikes; bracts and bracteolate persistent ending in spines, fruits utricles; seeds oblong.

Distribution: Occurs frequently along the edges of the forests and in shady places.

Therapeutic Usage:

Ethnogynaecology:

- Fresh root pulled out in one breath is stuck to the lock of the hair of the expecting mother, tied to the waist, or inserted into the vagina (if necessary) for the treatment of delayed delivery.
- It is claimed that the above practice is equally good for bringing out the dead child from the womb.

Ethnopharmacology:

- Root juice is given for mental disorders.
- About 2 gm of leaves and root powder mixed with black pepper powder is given to relieve asthma.
- Tribal ladies bite and brush their teeth with 365 pieces of stem during the 'Teej-Panchami' festival to purify their bodies from evils that happen within the year.
- 10-12 pieces of roots, tied to a string are hung around the goiter-affected neck till cure.

• Root extract is given orally in malarial fever thrice a day for 2-5 days. Root paste is applied on the body twice a day for 5-7 days.

FAMILY ANACARDIACEAE

Introduction

The Anacardiaceae family consists of several species of tropical and subtropical flowering plants. The genera Anacardium and Toxicodendron are two of the most well-known members of this family, respectively, because of their association with cashew nuts and poison ivy [29]. Some members of the Anacardiaceae family have a long history of folk medical usage. For instance, the cashew tree leaves and bark have been used in traditional medicine for centuries due to their purported anti-inflammatory and antibacterial effects [30]. The Anacardiaceae family is also highly relevant in ethnobotany, especially for indigenous peoples. Leaves stem bark, and roots, among other plant parts, are used for everything from treating medical conditions to performing religious rites [31]. The use of these plants has also advanced the field of ethnoveterinary medicine. Using regionally held expertise, traditional healers and livestock keepers have relied on extracts and medicines from specific Anacardiaceae species to treat and prevent diseases in domestic animals [32]. Some species of this plant family, like poison ivy, can be harmful or cause skin irritation, so it is essential to proceed with caution when working with them. A thorough study of the specific species and their qualities is essential to guarantee these plants' safe and efficient use for medicinal, ethnobotanical, or ethnoveterinary purposes. The plants belonging to the Anacardiaceae family exhibit a wide range of applications, encompassing both gastronomic pleasures and traditional medicinal practices. Although they have the potential to provide health advantages, it is imperative to utilize them with caution and consult healthcare specialists or experts for help when contemplating their medicinal uses.

• *Mangifera indica* **L.**

Local Name: Aam

Botanical Characteristic: The mango tree is a tropical fruit tree renowned for its delectable, succulent, and saccharine mango fruits.

Distribution: Cultivated

Therapeutic Usage:

Ethnopharmacology:

- Mango is not only a widely consumed tropical fruit but also a valuable provider of vital nutrients such as vitamin C, vitamin A, and dietary fiber. This fruit is consumed in its raw state and included in smoothies, salads, and various culinary preparations. Mango has a historical record of being utilized in traditional medicine, specifically in Ayurveda, because of its possible digestive and diuretic qualities.
- The Mango tree's foliage, outer covering, and reproductive units have been utilized in customary therapeutic practices.

- ***Buchanania lanzan*** Spreng.

Local Name: Kath Bhilwa

Botanical Characteristics: Medium-sized trees; stem with rough bark; leaves alternate, simple, broadly oblong, pubescent; flowers in axillary and terminal panicles, greenish-white; fruits drupes; seeds bony, two valved.

Distribution: Common in dry deciduous forests.

Therapeutic Usage:

Ethnopharmacology:

- Tribals prescribe stem bark decoction for washing septic wounds.
- They apply a mixture of gum and cow milk to alleviate rheumatic pain.
- A paste of cotyledons is applied to glandular pains.

Ethnoveterinary:

Tribals and rural people give leaf decoction in blood dysentery in cattle.

- ***Semecarpus anacardium*** Linn. f.

Local Name: Bhailwa

Botanical Characteristics: Medium-sized trees; leaves alternate, simple, large, ovate-oblong; flowers in terminal panicles; fruits drupes, obliquely ovoid, black when ripe, situated on yellow fleshy receptacle.

Distribution: Common in forest areas.

Therapeutic Usage:

Ethnogynaecology:

• Root decoction with long pepper paste is advised to women as a contraceptive.

Ethnopharmacology:

• Seed oil is applied on rheumatic swellings.
• Secretion of the pericarp is applied over the cuts and cracks of feet till cure.
• Nut oil massage is recommended for gout pain.

FAMILY APIACEAE

Introduction

• The Apiaceae family includes many different flowering plants easily distinguished by their hollow stems, complex leaves, and umbels of miniature flowers. It is one of Earth's most numerous plant families, with over 3,700 known species distributed among 434 genera [33]. Several members of the Apiaceae family have long histories of usage in medicine. Parsley (*Petroselinum crispum*), fennel (Foeniculum vulgare), and coriander (*Coriandrum sativum*) are just a few examples of medicinal herbs that have been used for centuries [34]. Essential oils from some species can be used for medical purposes. The Apiaceae family is essential in the study of ethnobotany. Several of its constituents are widely used as culinary herbs to add unique flavors and fragrances to dishes worldwide. Herbs like dill (*Anethum graveolens*) are good examples. In addition, some animals play an essential role in the rituals and ceremonies of different cultures. Certain plants in the Apiaceae family have also been used in ethnoveterinary medicine. Using local, generationally-passed expertise, traditional veterinarians and farmers treat and prevent diseases in domestic animals by administering plant extracts and parts [35, 36]. Although many members of the Apiaceae family have culinary, medicinal, and ethnobotanical applications, it is essential to use it with caution when working with this family because some of its species are hazardous. For responsible use, precise species identification is required. Plants belonging to the Apiaceae family are important in culinary traditions and possess possible health advantages in traditional medicine. Although highly regarded for their adaptability, it is crucial to exercise prudence and seek guidance from healthcare experts or traditional healers when contemplating their utilization for a medical purpose to guarantee both safety and efficacy of the therapy.

• *Daucus carota* L.

Local Name: Gaajar

Botanical Characteristics: An annual herb with alternate compound leaves, white umbrella-shaped flowers, and fleshy conical tap roots extended up to 5-50 cm in length. The stem is hairy and topped with umbel, white flowers.

Distribution: Cultivated

Therapeutic Usage:

Ethnopharmacology:

• *Daucus carota* is a highly recognizable and extensively consumed vegetable on a global scale. Carrots are renowned for their vivid orange hue, delightful taste, and outstanding nutritious content. They contain a high amount of beta-carotene, a substance converted into vitamin A that is crucial for maintaining healthy eyesight and overall health. Carrots are a versatile culinary component, commonly utilized in salads, stews, soups, and as a popular raw snack.
• It has long been acknowledged in traditional medicine for its digestive advantages and its role as a dietary fiber source to promote gut health.

• *Apium graveolens* L.

Local Name: Ajwain patta

Botanical Characteristics: Herb leaves thin, ovate, aromatic, and pinnately divided, stem long, narrow, hollow, and crisp, roots, fibrous, flowers off-white, fruits ovate, seeds tiny ovoid shaped.

Distribution: Wild/ cultivated.

Therapeutic Usage:

Ethnopharmacology:

• It is a crisp and fragrant vegetable frequently appreciated for its subtle and savory taste. It is a low-calorie food that provides a significant amount of dietary fiber. It is extensively utilized in culinary contexts, especially in salads and soups, and as a crispy snack, frequently served with dips, and is recognized for its inherent diuretic qualities, which can assist in maintaining fluid equilibrium and stimulating urine output.
• It has historically been employed in traditional herbal therapy for its capacity to mitigate inflammation and decrease blood pressure. In addition, celery seeds are

used as a seasoning and are said to possess therapeutic qualities, including anti-inflammatory and digestive advantages.

• *Coriandrum sativum* L.

Local Name: Dhaniya

Botanical Characteristics: Herb, annual, leaves compound, lobed or unlobed, tap root system, 120 cm tall, stem hollow, 30-60 mm tall with fetid smell, leaves pinnately decompounds with aromatic smell, seeds small and provided with oil ducts.

Distribution: Cultivated.

Therapeutic Usage:

Ethnopharmacology:

- *Coriandrum sativum*, commonly referred to as Coriander is an herb highly esteemed for its versatile characteristics. The foliage of this plant, commonly referred to as cilantro, is extensively utilized in gastronomic preparations, imparting a vibrant and tangy taste to various global cuisines. Seeds of it, however, serve as a spice and are essential to several spice blends and curry recipes.
- It has long been acknowledged in traditional medicine for its inherent digestive and anti-inflammatory qualities. It is thought to enhance digestion and ease gastrointestinal discomfort.

• *Petroselinum crispum* (Mll.) Fuss

Local Name: Ajmoda

Botanical Characteristics: Herb, leaves bright, triangular and aromatic, flowers small, yellow-green, fruits oblong with thread-like ribs, seeds have a flat face.

Distribution: Cultivated

Therapeutic Usage:

Ethnopharmacology:

- *Petroselinum crispum*, often known as Parsley, is a highly esteemed herb for its culinary uses and potential therapeutic properties. It serves as a decorative element and intensifies the taste of various types of food, including salads and soups.

- Parsley is abundant in vitamins A, C, and K, along with folate, and can serve as a valuable supplement to a well-rounded diet. Parsley is acknowledged in traditional medicine for its diuretic characteristics, which can potentially assist in managing water retention and urinary tract health.

• *Foeniculum vulgare* **Mill**

Local Name: Saunf

Botanical Characteristics: Perennial herb, 3-5 ft tall, leaves feathery compound, aromatic, yellow-green, flowers being small, yellow in color, fruits being dry seeds, stem being erect, tap root system.

Distribution: Cultivated.

Therapeutic Usage:

Ethnopharmacology:

- It is a multifunctional plant valued for its fragrant seeds and crunchy, anise-flavored bulb. Fennel seeds are a prevalent spice and digestive aid, frequently ingested post-meals to alleviate indigestion and gas.
- The plant is utilized in gastronomic concoctions in Mediterranean and Indian culinary traditions. Fennel is reputed in traditional medicine to possess digestive and anti-inflammatory attributes and has been employed in certain cultures to treat digestive distress and menstrual ailments.

• *Centella asiatica* (Linn.) Urban.

Local Name: Brami Booti

Distribution: Wild

Botanical Characteristics: A prostrate creeping herb, rooting at the nodes; leaves long-petioled, several at each node, flowers small, pink, in clusters of umbels, sessile; fruits being 2 seeded, indehiscent, laterally compressed; seeds are brown oblong.

Distribution: Found in moist places, meadows, along irrigation canals.

Therapeutic Usage:

Ethnopharmacology:

- The leaf juice has a coolant effect so given as an antidote for heat strokes.

- Whole plant extract is given to cure madness and mental disorders.
- Decoction of plants is sometimes recommended to cure leprosy-like symptoms. Leaf paste is applied in dermatitis-affected parts.
- Fresh flowers are collected, and sweet syrup made from them is frequently taken as memory memory-enhancing tonic, especially in children.

- ***Trachyspermum ammi*** (Linn.) Sprague

Local Name: Ajwain

Botanical Characteristics: An erect annual, glabrous or minutely pubescent herb; leaves 2-3 pinnate, flowers white, in compound umbels; fruits ovoid, muricate, sub hispid.

Distribution: Cultivated throughout the area.

Therapeutic Usage:

Ethnogynaecology:

In dysmenorrheal like painful menstrual periods, tribal women consume the seeds with warm water twice a day till get relief.

Ethnopharmacology:

- Seeds, asafetida powder, and rock salt are swallowed with lukewarm water to relieve stomach pain and gastric disorders.
- Extract made from seeds is very commonly used to treat toothache and gargling from this liquid is recommended for sore throat.

FAMILY APOCYNACEAE

Introduction

The Apocynaceae family is a large and varied genus of flowering plants in tropical and subtropical regions. Typical characteristics shared by members of this plant family include a milky latex sap and opposing leaves [37, 38]. The Apocynaceae family contains numerous essential plants used in traditional medicine. The alkaloids in periwinkle (*Catharanthus roseus*) are effective in treating cancer. The oleander (*Nerium oleander*) also has chemicals that may have cardiac glycoside characteristics [39]. These herbs have been used for centuries to treat various illnesses, both traditionally and in modern medicine [40]. Numerous Apocynaceae species are significant to the cultures who use them, lending to their ethnobotanical value. Some plants are used for more than just their medical properties; they are also a source of colors, fibers, or ceremonial materials. Their significance in local traditions is unique to each location and community [41]. The Apocynaceae family of plants has also been used in traditional veterinary medicine. Humans have long relied on the knowledge and practice of traditional healers and livestock keepers to treat and prevent diseases in domestic animals. The diversity and widespread distribution of the Apocynaceae family in many environments underline the importance of this plant group to human culture and traditional practices [42 - 44]. It is essential to use it with caution when working with plants from this family, as many of the species have medicinal and ethnobotanical benefits. To use them safely and effectively, one must first correctly identify them and understand their features. The plants belonging to the Apocynaceae family serve as a prime example of the significance of comprehending the aesthetic appeal and potential therapeutic applications of plants. However, it is crucial to exercise caution and seek expert advice, particularly when handling poisonous species such as Oleander and Yellow Oleander.

• *Hemidesmus indicus* (Linn.) R. Br.

Local Name: Gorkatala

Botanical Characteristics: Shrubs; leaves opposite simple, elliptic-oblong; flowers in axillary cymes, sub-sessile, purple; fruits follicles; seeds flat, black.

Distribution: Common in shady and moist places.

Therapeutic Usage:

Ethnopharmacology:

• Tribals prescribe root paste for the cure of leukoderma.
• Roots are crushed with 4 pepper seeds and garlic. This mixture is taken twice a day to cure dyspepsia.
• Root is powdered and mixed with coffee or tea powder in equal proportion. Coffee or tea is prepared with it and drunk for blood purification.
• Tribals hang 2 pieces of root on the neck of babies for proper development of skull bones.

• *Nerium oleander* **L.**

Local Name: Lal Kaner

Botanical Characteristic: Oleander is an aesthetically pleasing decorative shrub, characterized by clusters of appealing and aromatic blooms in hues of pink, white, or red. Nevertheless, it is crucial to exercise utmost vigilance when handling this plant, as it has a significant toxicity level. The Oleander plant contains many poisonous chemicals, specifically oleandrin and oleandrigenin. Even consuming minimal quantities of the plant can result in significant health complications, such as cardiovascular troubles and gastrointestinal discomfort.

Distribution: Wild/ cultivated.

Therapeutic Usage:

Ethnopharmacology:

- Historically, it has been employed for external usage to address skin ailments. However, owing to its highly poisonous nature, it must never be consumed orally.

- *Vinca minor* **L.**

Local Name: Sadauli

Botanical Characteristic: The Common Periwinkle is an attractive low-growing plant with beautiful blue-purple flowers and shiny green leaves.

Distribution: Cultivated.

Therapeutic Usage:

Ethnopharmacology:

- In addition to its decorative allure, it has become a staple in traditional herbal treatment. The presence of alkaloids such as vincamine has attracted scientific attention because of their ability to enhance cognitive function. These alkaloids are being studied for their potential to enhance cognitive function and memory by positively impacting cerebral blood flow. Nevertheless, additional investigation and verification are necessary to utilize these chemicals in medicinal contexts.

- *Thevetia peruviana* **(L.) Lippold**

Local Name: Peeli Kaner

Botanical Characteristics: A shrub or small tree, up to 25 ft tall, leaves linear, alternate, flowers funnel-shaped, bright yellow or orange, fruits green when unripe, triangular, sap milky and toxic in nature.

Distribution: Cultivated.

Therapeutic Usage:

Ethnopharmacology:

• This is an aesthetically pleasing shrub that displays vivid yellow flowers. Like Oleander, this plant is extremely poisonous, and all its parts, including the seeds and sap, contain dangerous cardiac glycosides. Consuming any portion of the plant might be fatal.

• Conventional applications involve the application of a substance directly to the skin to treat various skin problems. Nevertheless, using utmost prudence when handling Yellow Oleander is crucial, and it must never be consumed.

• *Alstonia scholaris* (Linn) R. Br. in mem. Werm. Nat. Hist. s. 1:75. 1811.

Local Name: Ajan, Chatwan

Botanical Characteristics: Trees, evergreen, with grey bark, leaves in whorls 4-7; flowers greenish white, sweet-scented; fruits follicles in clusters; seeds linear, oblong with a tuft of hairs at each end.

Distribution: Found in moist regions, common on roadsides and outside the forests.

Therapeutic Usage:

Ethnopharmacology:

• For the easy expulsion of helminths from the intestine, the extract of stem bark along with long pepper (*Piper longum*)is given 5 spoonfuls twice a day for three days.
• Tribal people frequently apply latex of plants to cure caries of teeth.
• Root paste can be applied on tumors and cancerous overgrowths.
• Paste of stem bark is applied for chest and joint pains.
• 5 ml of stem bark decoction is given thrice a day for 3-5 days in prolonged fevers.

Ethnoveterinary:

Tribal and rural people prescribe latex with a decoction of black pepper (4:3) to cattle for the expulsion of intestinal worms.

• *Carissa opaca* Stapf. ex. Hains.

Local Name: Jangli karonda

Botanical Characteristics: Shrubs; stem with milky juice and forked thorns; leaves opposite, simple, elliptic ovate; flowers in axillary and terminal cymes, white; fruits berries, dark purple when ripe.

Distribution: Common in hedges in scrub jungles.

Therapeutic Usage:

Ethnopharmacology:

• Immature fruits and fresh flowers are taken orally in the morning to cure cough and cold for about 9 days.
• Tribals apply root bark paste on the ulcers of the human body.
• In the treatment of remittent fever, tribals prescribe stem decoction with black pepper paste in a 5:2 ratio.
• Rural people put 4-5 pieces of roots near their rooms' entrances to keep snakes and other harmful creatures away. Similarly, they grow plants near their houses to keep snakes away.

Ethnoveterinary:

Powder of dried roots is placed on wounded and wormed parts to expel them.

• *Catharanthus roseus* (Linn.) G. Don.

Local Name: Sadabahar

Botanical Characteristics: An herb or small shrub; leaves are opposite, entire, deep green, polished, obovate or oblong petiole, 2 glandular; flowers white or pink or reddish in color; follicles 2-3.5 cm long, narrowly cylindrical.

Distribution: Cultivated or found as an escape.

Therapeutic Usage:

Ethnopharmacology:

• For the sure cure of diabetes, 2-3 green and fresh leaves are chewed as such.
• Leaves juice is rubbed over the wasp-bitten parts.

• *Holarrhena antidysenterica* (Roth.) A. DC.

Local Name: Duddhi

Botanical Characteristics: Small, deciduous trees; leaves opposite, simple, ovate, elliptic; flowers in axillary and terminal corymbose cymes, white; fruits in pairs, long, cylindrical, follicles; seeds linear-oblong with deciduous comas.

Distribution: Very common in forest areas

Therapeutic Usage:

Ethnogynaecology:

A decoction made from stem bark is given to expecting women to cure or minimize labor problems.

Ethnopharmacology:

• As the name of the plant indicates, the powder of the plant is given in 1 spoonful twice a day for four days for the cure of dysentery.
• Decoction of stem bark with honey is frequently advised to patients with epithelial tumors.

• *Rauwolfia serpentina* (Linn.) Benth. ex-Kurz

Local Name: Nag bel

Botanical Characteristics: Under shrubs; leaves in whorls, simple, lanceolate; flowers in corymbose cymes, tubular, pinkish white; fruits berries, globose; seeds solitary, ovoid.

Distribution: Commonly found wild under shades, also cultivated.

Therapeutic Uses:

Ethnopharmacology:

• 1 spoonful of root powder is given twice daily for three days to treat fever. Root extract is also given two times a day for three days.
• The root is crushed and the paste is applied externally on the snake-bitten area.
• Root paste 10 gm is given for hypertension and related problems. It also lowers the blood pressure.
• Root bark with honey is given to cure insanity.

FAMILY ARACEAE

Introduction

The Araceae family, or the "arum family," is a large group of flowering plants in warm and subtropical areas. These plants stand out in different settings because of their distinctive leaves and flowering structures. Some members of the Araceae family have been used for centuries as medicines [45]. For example, the root of the wild taro plant (*Colocasia esculenta*) is thought to have anti-inflammatory and pain-relieving properties, and parts of the elephant ear plant (*Alocasia spp.*) have been used for what were thought to be their healing qualities [46]. Regarding ethnobotany, Araceae plants are essential in many countries [47]. In addition to being possible medicines, they are often used as food items, with the taro plant (*Colocasia esculenta* and *Xanthosoma spp.*) being a primary food source in many places [48]. Also, some Araceae plants are used in ceremonies and rituals in indigenous societies. Some Araceae plants have been used in traditional veterinary treatment, but this is not as common with this plant family as with other plant families [49, 50]. Local knowledge has been used to use certain parts of plants or plant extracts to help sick animals. Some Araceae species have medicinal uses, others are used in regional cuisines, and others have cultural meanings. This shows their importance to many parts of human culture and traditional practices.

- *Colocasia esculenta* (L.) Schott.

Local Name: Ghuiyaan

Botanical Characteristic: It is a tuberous plant with an underground stem called a corm. It is a tropical root vegetable known for its starchy and nutty taste.

Distribution: Cultivated.

Therapeutic Usage:

Ethnopharmacology:

- This cuisine is commonly consumed in numerous tropical and subtropical areas, including Asia, the Pacific Islands, and certain sections of Africa.
- It is utilized in many gastronomic preparations, ranging from savory stews to delectable desserts. In order to neutralize naturally present toxins, such as calcium oxalate crystals, which can create a burning sensation if ingested raw, it is necessary to cook it completely. When well cooked, it serves as a significant

reservoir of carbohydrates and offers vital nutritional support in numerous societies.

• *Syngonium podophyllum* Schott

Local Name: Gusoot

Botanical Characteristics: Evergreen climbing vein grows up to 3-6' long, leaves are lobed and exude a white sap when broken. Stem also exudes a white sap when broken, flowers borne on a spadix.

Distribution: Cultivated.

Therapeutic Usage:

Ethnopharmacology:

• It is a widely favored indoor plant due to its aesthetically pleasing leaves. The leaves of this plant are often shaped like an arrowhead and display various colors and patterns. These plants are low-maintenance, making them a popular option for indoor ornamentation. They are valued for their ability to filter the air and are frequently used to improve indoor air quality.
• Leaves are often used in wound healing and to treat bacterial infections.

• *Epipremnum aureum* (Linden & Andre) G.S. Bunting

Local Name: Money plant

Botanical Characteristics: A climber or creeper with stems growing up to 12 cm long, leaves showing a yellow marble pattern, young plants have waxy, heart-shaped green leaves, stems up to 4 cm diameter, climbing using aerial roots.

Distribution: Cultivated.

Therapeutic Usage:

Ethnopharmacology:

• It is a popular houseplant admired for its cascading tendrils and foliage shaped like hearts. This plant is versatile and requires minimal upkeep, making it suited for both novice and seasoned gardeners. Devil's Ivy is renowned for its air-purifying capabilities and is regarded as a highly effective option for improving indoor air quality. Its durability and visual attractiveness have established it as a fundamental element in both indoor gardening and interior design.
• Leaves are used to treat fungal and bacterial infections in patients.

• Rural and ethnic people have been using root decoction to manage cancerous growths.

• *Acorus calamus* (L.) Hook. f.

Local Name: Papari

Botanical Characteristics: Semi-aquatic aromatic herbs with creeping root-stocks; leaves simple, sessile, linear with wavy margins; flowers in spadix; fruits berries.

Distribution: Common wildly in marshy areas, also cultivated

Therapeutic Usage:

Ethnopharmacology:

• To treat chicken pox, root paste with *Piper longum* (Pipal) is given thrice a day for five days.
• Small pieces of rhizomes are chewed for an instant cure of sore throat, bronchitis, cold, and cough.
• The rhizome is roasted and powdered. A pinch of this powder is given for cold and fever.
• Tribals give root paste with a paste of 7 long peppers to children with fever caused by supernatural powers as per their beliefs.

FAMILY ASCLEPIADACEAE

Introduction

The Asclepiadaceae family, also called the milkweed family, comprises many flowering plants in warm and subtropical areas [51, 52]. These plants are found all over the world. They have unique flower shapes and milky latex sap. Many plants in the Asclepiadaceae family have been used traditionally as medicines. The tropical milkweed (*Asclepias curassavica*) is known for its possible medical uses [53]. Different parts of species in this family have recently gotten attention from the medicine industry [54]. Asclepiadaceae plants have cultural meaning in many societies from an ethnobotanical point of view. In addition to being used as medicines, some species are important in culture and rituals. They play essential roles in religious ceremonies and traditional rites, varying from region to region and community [55]. Even though the Asclepiadaceae family is not as often used in traditional veterinary treatment as some others, certain species of Asclepiadaceae have been used in this way [56]. Indigenous knowledge has guided using certain plant parts or extracts to help household animals with health problems [57]. The variety of the Asclepiadaceae family shows how important it is to human culture and traditional knowledge systems. Some family members have medicinal uses, others are used in cultural practices, and a few might be used in ethnoveterinary applications [58]. Still, for these plants to be used safely and effectively in traditional ways, they must be correctly identified, and their possible toxicity must be understood.

• *Asclepias syriaca* **L.**

Local Name: Duddhi.

Botanical Characteristics: A perennial herb with an erect stem that can grow up to 5 ft tall, leaves oblong, opposite, flowers are pink to white, with umbel inflorescence, fruits (pods) are about 10 cm long.

Distribution: Wild

Therapeutic Usage:

Ethnopharmacology:

• It is an indigenous plant of North America highly valued for its crucial function in assisting Monarch butterflies. It is a vital host plant for Monarch caterpillars, offering nourishment and protection throughout their larval phase. The plant exhibits clusters of pink to lavender flowers that have a pleasant, fragrant fragrance. In the past, Indigenous communities used the plant's durable fibers to create cordage and textiles, which greatly contributed to its value as a resource. Nevertheless, it is crucial to acknowledge that Common Milkweed possesses milky latex that can be poisonous when consumed. Consequently, it is unsuitable for culinary or medicinal use and should be treated cautiously.
• Root decoction has been recommended to get relief from pain due to rheumatism and backache.
• The plant also has anti-oxidant properties and is useful in kidney and urinary dysfunctions.

• *Ceropegia woodii* **Schltr.**

Local Name: Dil Bel

Botanical Characteristics: Tender perennial plant, leaves heart-shaped, succulent, flowers tubular, lantern-shaped, white to pale magenta in color, stem long, wiry, roots tuberous.

Distribution: Cultivated.

Therapeutic Usage:

Ethnopharmacology:

- It is an endearing and well-cherished succulent houseplant. The plant is renowned for its hanging vines covered in petite, heart-shaped foliage. This plant is highly valued for its decorative attributes and is commonly recognized in suspended containers or as a cascading indoor gardener. It necessitates ample, indirect illumination and soil that drains effectively.
- Leaves decoction is recommended for combating anxiety in stressful conditions.
- The plant also has allergic and anti-bacterial properties.

• *Hoya carnosa* **(Lf) R.Br.**

Local Name: Mome paudha

Botanical Characteristic: It is a well-liked indoor plant admired for its glossy, star-shaped blossoms and appealing, dense foliage. This plant is well esteemed for its decorative attractiveness and is frequently grown indoors or suspended.

Distribution: Cultivated.

Therapeutic Usage:

Ethnopharmacology:

- The plant is widely called the Hindu Rope Plant because of its distinctively knotted vines. Wax plants are recognized for their durability and generally undemanding care needs, which consist of bright, indirect light and soil that drains well. Due to their unique appearance and occasional emission of fragrant flowers, they are highly preferred for indoor gardening and decorative arrangements.
- Leaves are proven to have anti-diabetic properties and can cure urinary disorders as well.

• *Asclepias curassavica* **L.**

Local Name: Kaktund

Botanical Characteristic: Tropical Milkweed is a visually appealing and decorative variety of milkweed, characterized by its vivid red and yellow flowers. It strongly appeals to pollinators, notably butterflies, and is highly favored by gardeners for its crucial function in providing sustenance to these insects.

Distribution: Wild/Cultivated

Therapeutic Usage:

Ethnopharmacology:

Nevertheless, the utilization of Tropical Milkweed may be a topic of contention in areas where the preservation of Monarch butterfly populations is a priority. Although this species offers nectar and nourishment, several localities refrain from cultivating it to save the Monarchs' innate migratory patterns. Tropical Milkweed is an aesthetically pleasing plant that enhances gardens and is highly valued for its ecological importance in providing support to pollinators.

• *Calotropis gigantea* (Linn.) R. Br. in Ait.

Local Name: Akauwa

Botanical Characteristics: Shrubs; leaves are opposite, simple, subsessile, ovate-oblong; flowers in umbellate cymes, white; fruits follicles; seeds being long, ovate, smooth, with long comas.

Distribution: Common in forests and wastelands

Therapeutic Usage:

Ethnopharmacology:

- Tribals apply fomented leaves on the abdomen for relief against abdominal pain.
- Rural and tribal people apply fomented fresh leaves to cure swelling due to sprain.
- They apply latex in the treatment of toothache.
- If any thorns or spines are inserted deeply in any part of the body (which cannot be taken out by mechanical means); rural people put 2-3 drops of latex on the affected part and tie them with fomented leaves of said plant overnight. It helps to bring out the inserted spine without any pain.
- Fresh leaves warmed in mustard oil are applied on rheumatic pain-affected parts as a bandage.
- Powder of root bark is given to cure dysentery and indigestion.
- In jaundice, 1 teaspoon of root bark powder is given two times a day for three days.

FAMILY ASTERACEAE

Introduction

The Asteraceae family, also called the aster or daisy family, is one of the world's most extensive and diverse flowering plants. It includes more than 1,900 genera and about 32,000 kinds of well-known plants, like sunflowers, daisies, and chrysanthemums [59]. Many of its members have been used as medicine in both old and new ways. For example, arnica (*Arnica montana*) has been used for its possible anti-inflammatory effects, and chamomile (*Matricaria chamomilla*) is well-known for its soothing effects [60]. These plants and many others have been used to make herbal medicines and pharmaceuticals. Ethnobotanically, plants in the Asteraceae family have cultural meaning in many countries. Some species are used in cooking and medicine [61]. For example, safflower (*Carthamus tinctorius*) is a spice and flavoring agent [62]. Also, some Asteraceae plants are used in ceremonies and rituals in different countries, which shows how deeply they are tied to local traditions [63]. Some members of the Asteraceae family have been used in the field of ethnoveterinary medicine. Using their knowledge and experience, traditional healers and livestock owners have used specific parts of plants or extracts from plants to treat and avoid diseases in domestic animals [64]. Many of these species have medicinal uses, can be used in cooking, or have cultural meaning. This shows how important the family is to different parts of human culture and traditional practices. To make sure that these plants can be used safely and effectively in different ways, it is essential to identify them and know if they are poisonous correctly.

• *Helianthus annuus* **L.**

Local Name: Suryamukhi

Botanical Characteristics: Sunflowers, formally referred to as Helianthus annuus, are renowned for their vibrant yellow petals and their distinctive capacity to track the sun's trajectory throughout the day, a process termed heliotropism.

Distribution: They are renowned for their decorative aesthetics and are frequently cultivated in gardens for their visual allure. In addition to their visual appeal, sunflowers have both gastronomic and nutritional benefits.

Therapeutic Usage:

Ethnopharmacology:

• The seeds, usually known as sunflower seeds, are a popular snack and serve as a source of sunflower oil, utilized in cooking and numerous culinary products. Sunflower oil is widely recognized for its subtle taste and elevated smoke point, making it a versatile culinary option. Leaves have properties to get relief from soar throat, cough and cold, malaria, and pulmonary infections.
• Seed oil is very useful in minimizing the risks of heart troubles.

• ***Taraxacum officinale* Weber**

Local Name: Singhparni

Botanical Characteristics: They are easily identifiable wildflowers characterized by their vibrant yellow blossoms and distinctive spherical seed heads. These plants are commonly regarded as weeds; however, they possess a lengthy historical background of being utilized in cooking and medicine.

Distribution: Cultivated.

Therapeutic Usage:

Ethnopharmacology:

• Its leaves are edible and commonly used in salads, imparting a little bitter taste. The roots have been utilized in traditional herbal therapy due to their putative hepatoprotective and digestive properties. Utilizing dandelion root tea, tinctures, and supplements is a prevalent method to leverage these potential health benefits. The plant's adaptability in both culinary and medicinal contexts renders it a highly valued and sought-after natural asset.

• ***Calendula officinalis* L.**

Local Name: English Genda

Botanical Characteristic: Annual or perennial herb with waxy, smooth or glandular stems, leaves simple, alternate, sessile, long, with entire margins, stem may be spiny, flower bright yellow, leaves are aromatic, fruits dry thorny.

Distribution: Cultivated.

Therapeutic Usage:

Ethnopharmacology:

- It is admired for its beautiful, daisy-like flowers that display a spectrum of vivid hues including orange and yellow. These vibrant flowers are widely favored in gardens for their aesthetic appeal. Marigold flowers are acknowledged in traditional herbal therapy for their potential to soothe the skin and reduce inflammation.
- Calendula-infused oils and creams mitigate skin irritations, such as rashes and small burns. The plant's mild yet potent characteristics have established it as a fundamental ingredient in organic skincare products. It serves as a natural source of yellow and orange colors, commonly utilized in food and textiles, in addition to its decorative and therapeutic applications.

- *Chrysanthemum indicum* **L.**

Local Name: Guldaudi

Botanical Characteristics: Chrysanthemums are blooming plants that are part of the Chrysanthemum genus. They are well-known for their colorful, daisy-like flowers. They are highly valued in gardens, floral compositions, and cultural festivities. Chrysanthemums exhibit diverse colors, each of which carries its own distinct symbolic significance.

Distribution: Cultivated.

Therapeutic Usage:

Ethnopharmacology:

- Chrysanthemum holds significant cultural importance in numerous Asian societies, symbolizing longevity and playing a crucial role in diverse ceremonies and festivals. The flowers have served as a source of relief from inflammation, and microbial infections and as a good anti-oxidant.
- The decoction made from fresh flowers is recommended for hypertension and respiratory disorders.

- *Elephantopus scaber* L.

Local Name: Bishari

Botanical Characteristics: Herbs; short rootstocks with stout fibrous roots; leaves radical, obovate-oblong, serrate; flowers in the head, bisexual; fruits truncate, brown, ribbed, covered by mucilage.

Distribution: Common as a weed in shady places.

Therapeutic Usage:

Ethnopharmacology:

- Dried root paste is mixed with mustard oil and applied to the eczema-affected area.
- For the treatment of fever due to measles in children, fresh root extract is prescribed.
- Crushed leaves and roots boiled in water and given to stop vomiting.
- Root paste is very useful in all poisonous bites or nails of wild animals.
- Roots are used to treat filarial swellings.

• *Vernonia anthelmintica* (Linn.) Willd.

Local Name: Kalijiri

Botanical Characteristics: Herbs; annual, erect up to 80 cm in height; leaves lanceolate, elliptic-lanceolate; heads corymbs; peduncles with linear bract near the top; fruits cypsela, oblong, angular.

Distribution: Common in waste places, not found in water-logged conditions.

Therapeutic Usage:

Ethnopharmacology:

- Seeds are used as as anthelmintic.
- Seed powder is found very effective against threadworms.
- Fresh leaf juice is given to cure piles and dysentery.
- Seeds are prescribed for leucoderma.

FAMILY BIGNONIACEAE

Introduction

The Bignoniaceae family, also called "bignonias" or "trumpet creepers," is a large group of flowering plants with showy flowers that often resemble trumpets. This family has about 110 genera and more than 800 species, most of which live in warm or subtropical areas [65, 66]. Some plants in the Bignoniaceae family have been used for medicine because of their possible healing qualities. For example, the bark of the Tabebuia tree (*Tabebuia impetiginosa*), also called pau d'arco, has been used in traditional medicine in Central and South America because it is thought to have anti-inflammatory and antimicrobial benefits [67]. From an ethnobotanical point of view, Bignoniaceae plants are essential to different cultures. In addition to being used as medicine, some species are used for cultural and religious reasons. Their importance can change from place to place, depending on how they are used in local traditions [68]. Even though the Bignoniaceae family of plants is not as often linked to ethnoveterinary practices as some other plant families, some species may have been used in traditional veterinary treatment [58]. The diversity of the Bignoniaceae family shows its role in human culture and traditional knowledge systems [69]. For example, some members may be helpful in medicine, have cultural meaning, or have a limited role in ethnoveterinary practices. When using these plants for different purposes, one must be careful and ensure that the species are correctly identified and that you know their qualities [70, 71].

• *Oroxylum indicum* (Linn.) Vent., For.

Local Name: Sauna

Botanical Characteristics: Small trees; leaves opposite, pinnately compound; flowers in terminal racemes, large, bell-shaped, lobed, purplish; fruits capsules; flat, long, sword-like; seeds broadly ellipsoid, hyaline, silvery winged.

Distribution: Almost throughout forest areas, common in open forest areas, roadsides, and villages.

Therapeutic Usage:

Ethnogynaecology:

Stem bark decoction is recommended for women with dysmenorrhea.

Ethnopharmacology:

• Tender fruits are used to cure stomach aches.
• Tribals use root bark paste with the paste of long pepper (3:2) to cure rheumatic swellings.
• Root bark paste is given thrice daily for two days to cure diarrhea and dysentery.
• Stem bark juice with cow's milk is taken orally for 10 days to set bone fractures.

Ethnoveterinary:

Stem bark paste is applied externally over the fractured bones in cattle.

FAMILY BOMBACACEAE

Introduction

• The Bombacaceae family, usually known as the baobab or kapok family, includes several flowering plants worldwide, particularly in warm and subtropical regions [72]. Their large trunks and colorful blossoms distinguish these plants. The Bombacaceae family has been used in medicine for its therapeutic properties. The baobab tree (*Adansonia spp.*) is noted for its many uses, including eating its healthy fruits and leaves [73]. These herbs have been utilized in traditional medicine for many ailments. Many cultures value Bombacaceae plants ethnobotanically. In addition to medication, they are utilized in cooking. Baobabs are utilized for their flavor and health advantages [74, 75]. Some species are used in other nations' ceremonies and rituals, demonstrating their ties to local customs. Some Bombacaceae plants may have been employed in traditional veterinary therapy, albeit they are not as typically associated with ethnoveterinary practices [76, 77]. Based on local knowledge, plant parts or products cure animal health issues. Bombacaceae family members may be medicinal, edible, or culturally significant [78]. This shows how important the family is in many cultures and traditions. Identifying and knowing if certain plants are harmful is crucial to using them safely and efficiently. The Bombacaceae family comprises trees that possess not only botanical intrigue but also cultural and economic importance. They have made significant contributions to the sustenance and customs of several societies, rendering them remarkable in both botanical and human spheres.

• *Adansonia digitata* **L.**

Local Name: Gorakh Imli

Botanical Characteristic: Large tree, deciduous, tall up to 80 ft, bark smooth, leaves are palmately compound and made of 5-7 leaflets, flowers are creamy white, large and pendulous, fruits are egg-shaped and covered with yellowish hairs.

Distribution: Cultivated.

Therapeutic Usage:

Ethnopharmacology:

- It is recommended for its colossal trunk and distinctive white blossoms, earning it the nickname "Tree of Life." The tree is renowned for its multifaceted applications, as virtually all its components possess significant value.
- The fruit pulp is abundant in vitamin C and is employed in beverages and traditional treatments. The excavated tree trunks have functioned as protective structures, repositories, and even confinement facilities.
- Seeds are recommended for analgesic, anti-oxidant, and anti-microbial properties.

- *Litsia speciosa* **(ASt.-Hil., A.Juss. & Cambess.)**

Local Name: Resham Rui

Botanical Characteristic: It is a shrub or tree, leaves are shiny, the flowers are small, 15 mm in diameter, yellow-brown in color, and the fruits are oblate.

Distribution: Wild.

Therapeutic Usage:

Ethnopharmacology:

- It is renowned for its remarkable aesthetics, characterized by towering trunks adorned with thorny protrusions and vibrant pink or white blossoms. "Silk Floss" denotes the downy and fibrous material that envelops the tree's seeds. These fibers have been utilized for several applications, such as filling pillows and cushions. The stem bark has anti-bacterial, anti-inflammatory, and antipyretic properties.
- Seed oil has extensively been used as hepatoprotective and as a good anti-oxidant.

- *Ceiba pentandra* **(L.) Gartner.**

Local Name: Safed Semal

Botanical Characteristic: It is predominantly found in tropical rainforests and boasts an impressive stature, making it one of the tallest trees on the planet. The tree's towering, vertical trunk and distinctive branch pattern make it readily identifiable. The tree yields sizable pods containing kapok, fluffy, cotton-like fiber. These fibers possess a low weight and the ability to float, making them suitable for conventional applications such as life jackets, pillows, and insulation. The Kapok tree is culturally significant in numerous locations, sometimes regarded as sacred or esteemed.

Distribution: Wild

Therapeutic Usage:

Ethnopharmacology:

- Its leaves and bark have anti-acne properties to get rid of Acne. Apart from it, it also has anti-microbial properties and is recommended for Chickenpox and other skin problems.
- It has an anti-aging capability to minimize aging effects on the skin.

• *Pachira aquatica* **Aubl.**

Local Name: Money plant

Botanical Characteristic: Medium-sized tree, 15-20 mt height, with whorled branching, evergreen, leaves glossy, palmate with 5 petals, thick roots, flowers green-yellow.

Distribution: Cultivated.

Therapeutic Usage:

Ethnopharmacology:

- It is renowned for its unique intertwined stem and fan-shaped leaves. It is frequently employed in Fang Shui practices to draw in favorable energy and abundance. Stem bark is used to cure stomach problems, and to treat anemia in children.
- Leaves are used by tribals to control high BP and the conditions of fatigue and debility in women.

• *Bombax ceiba* L.

Local Name: Simra, Semargulla

Botanical Characteristics: Tall deciduous trees forming spreading crowns; stems with grey bark; leaves digitately compound, 1-7 foliate, acuminate, entire, glabrous; flowers large, bright red, solitary or in clusters, thick, cup-shaped, densely silky; fruits woody capsule; seeds black, round, concealed with long dense white silky hairs.

Distribution: Common in forests and on roadsides.

Therapeutic Usage:

Ethnopharmacology:

- Tribals give 5gm of stem exudates to children to cure diarrhea.
- Tribal and rural people apply flower paste on smallpox wounds.
- Fruit paste is applied on leprotic wounds.
- Stem bark paste with salt is applied on boils.
- Sepals of the fallen flowers are cooked as vegetables and advised to the women complaining of back ache and dysmenorrheal pain.

Ethnoveterinary:

- Ethnic persons give stem bark decoction to cure diarrhea in the cattle.

FAMILY BORAGINACEAE

Introduction

The Boraginaceae family, or borage or forget-me-not family, is a varied worldwide flowering plant family. Boraginaceae members vary in form, habitat, and ecological roles, with over 100 genera and 2,000 species [79, 80]. The family inhabits arid deserts, alpine meadows, and temperate climates. These plants have bristly alternating leaves and five-petaled, tubular, or funnel-shaped blooms [81]. Borage, comfrey, and heliotrope are well-known genera in this family. Boraginaceae plants are essential in traditional herbal therapy. Borage is utilized for its anti-inflammatory and calming effects and comfort for wound healing [82]. The family's rich flora has many therapeutic uses in many cultures. Many cultures value Boraginaceae species ethnobotanically. Some members are used in cooking as well as medicine [83]. Borage flowers are garnished in salads and drinks. Boraginaceae plants may also be used in rituals and ceremonies, reflecting local traditions [84]. Boraginaceae species may have been utilized in traditional veterinary treatment; however, the family is not as well-known for ethnoveterinary practices [85]. Local knowledge guides the use of plant components or extracts for animal health. The Boraginaceae family's extensive distribution, rich flora, and possible uses in health, culinary arts, and culture make it essential in human culture and traditional knowledge systems [86]. However, precise plant identification and toxicity knowledge are essential for safe and effective plant use in diverse applications.

• *Cordia dichotoma* Forst.f.

Local Name: Lasaura

Botanical Characteristics: Medium-sized deciduous trees with grey or brown bark; leaves ovate, elliptic to suborbicular; flowers white in lax, terminal, or axillary cymes; drupes ovoid, apiculate.

Distribution: Found as small trees in forests, villages, and roadsides.

Therapeutic Usage:

Ethnogynaecology:

To cure dysmenorrhea, stem bark along with the bark of *Ficus racemosa* and prop roots of *Ficus benghalensis* are taken in a 2:1:1 ratio to make an extract, and 20-30 ml is given twice a day till cure.

Ethnopharmacology:

• Tribal and rural people very often consume ripe fruits for purgative actions.
• Tender fruits are cooked as vegetables and prescribed as laxatives to constipation patients.
• Leaf decoction with salt is used in common coughs and cold.
• For the cure of diabetes, 10 ml aqueous extract of aerial roots along with salt is given once a day in the morning for one month.

FAMILY CAESALPINIACEAE

Introduction

The senna or cassia family, the Caesalpiniaceae, is a varied collection of flowering plants found worldwide, mostly in tropical and subtropical environments [87]. These plants are ecologically, medicinally, and culturally crucial due to their pinnately complex leaves and unique flower forms. Traditional medicine uses various Caesalpiniaceae species. *Senna alexandrina* and *Cassia spp.* are known for their laxative properties [88, 89]. Traditional medicine relies on these plants for gastrointestinal and other health conditions. Caesalpiniaceae plants are culturally significant in many communities, while some species have culinary uses. Tamarind (*Tamarindus indica*) is utilized in cuisine worldwide for its sweet and acidic fruit. Caesalpiniaceae plants also have ritual and ceremonial significance in diverse societies, demonstrating their profound ties to local customs [90, 91]. Caesalpiniaceae may be used in ethnoveterinary practices. Indigenous knowledge guides domestic animal disease treatment and prevention with plant parts or extracts [92]. The richness of the Caesalpiniaceae family, with medicinal, gastronomic, and cultural uses, emphasizes its relevance in human culture and traditional practices [93]. As with any plant use, precise identification and toxicity knowledge are essential for safe and effective use in many applications.

• *Caesalpinia crista* L.

Local Name: Khaja

Botanical Characteristics: Climbers; stem with prickles; leaves pinnately compound; flowers in dense racemes, yellow; fruits pods; seeds oblong, lead-colored.

Distribution: Common in hedges.

Therapeutic Usage:

Ethnogynaecology:

Tribals give leaf powder with a decoction of long pepper (3:2) to pregnant ladies as a general tonic.

Ethnopharmacology:

• Dried seed powder is given for dyspepsia.
• Dried seed powder is also prescribed for malarial fever. Young and neem leaves are roasted, made into powder, and given thrice daily.
• Seed oil massage is administered for facial paralysis.
• For stomach pain, seed paste is given for one day.

Ethnoveterinary:

Tribal and rural people give root decoction (about 50 ml) with a paste of 20 black peppers to their cattle for quick expulsion of the placenta after delivery.

• *Cassia fistula* L.

Local Name: Sinara

Botanical Characteristics: Medium-sized trees; leaves pinnately compound; flowers in racemes, yellow; fruits pod, long, cylindrical; seeds flat, black, embedded in sweetish pulp.

Distribution: Common in forests and on roadsides.

Therapeutic Usage:

Ethnopharmacology:

- For treatment of jaundice, seed powder, and fruit pulp are used.
- For septic wounds, stem bark decoction is used.
- Tribals use sweet pulp as a laxative to cure constipation.
- Dried root powder is given once a day for 30 days to treat liquor addiction.
- Ethnic persons apply a decoction of stem-bark on leprosy affected part till cured.
- Ash of fruits along with honey is given orally for whooping cough.

Ethnoveterinary:

- Tribal people use fruit powder with mustard oil and turmeric powder (4:1:2) for their cattle in intestinal worms.
- Pulp of fruit is given to cattle in indigestion.

- ***Cassia tora*** L.

Local Name: Chakwar

Botanical Characteristics: Herbs, and leaves are pinnately compound; flowers in terminal racemes, yellow; fruits pods, sub-terete; seeds being brown.

Distribution: Abundant as a weed in waste places.

Therapeutic Usage:

Ethnopharmacology:

- Tribals apply a paste of root bark with egg albumin for fast recovery in bone fractures.
- Paste of fresh leaves in mustard oil is applied locally on eczema-affected areas.
- In fresh wounds and cuts, tribals apply green leaf juice to avoid infection and for quick healing.
- Dried seed powder with mustard oil is found to be adequate to control ringworm.
- 5 ml of seed decoction is given three times a day for three days for malarial fever.

• *Cassia occidentalis* L.

Local Name: Kasondha

Botanical Characteristics: Under shrubs, diffused; leaves are pinnately compound; leaflets in pairs; flowers in axillary corymbose racemes, yellow with reddish veins; fruits pods; seeds compressed, black.

Distribution: Grows in waste places and along the roadsides.

Therapeutic Usage:

Ethnopharmacology:

- For diabetes treatment, tribals prescribe powder of dried root bark with seed powder of *Syzygium cuminii* (3:2), two teaspoonfuls of this powder is given with honey once a day for one month.
- Root paste with lime (3:1) is used for ringworms.
- Roasted fruits and seeds are crushed. One gm of this powder is mixed with rape seed oil and applied externally on scabies.

FAMILY CAPPARIDACEAE

Introduction

The mustard family, Capparidaceae, is a widespread collection of flowering plants worldwide. This family of 40 genera and 700 species has many shapes and adaptations to different ecological niches [94, 95]. The Capparidaceae family is particularly abundant in arid and semi-arid environments, from deserts to tropical rainforests. The plants in this family have pinnately complex leaves, four- or six-petal blooms, and fragrant glands. Traditional medicine uses Capparidaceae family members. Kaper *(Capparis spinosa)* has been used for its anti-inflammatory and antioxidant effects [96]. These herbs have helped traditional medicine treat many illnesses. Different cultures value Capparidaceae plants ethnobotanically. Some species are used in cooking and medicine: caper, a famous condiment from pickled flower buds and fruits [97, 98]. In addition, some Capparidaceae species are associated with rituals and ceremonies in different civilizations. Capparidaceae plants may have been utilized in traditional veterinary care, albeit not as well-known as other plant families. Local knowledge guides using plant components or extracts for animal health [99]. The Capparidaceae family's extensive distribution, diversified flora, and possible uses in health, culinary arts, and culture make it essential in human culture and traditional knowledge systems [100]. Precise plant identification and toxicity knowledge are necessary to safely and effectively use these plants in various applications.

• *Capparis zeylanica* L.

Local Name: Zakhambela, Harralura

Botanical Characteristics: Climbing shrubs; stems with recurved thorns; leaves alternate, simple, elliptic-oblong, tomentose; flowers in super axillary racemes, white or pink; fruits berries, subglobose, red when ripe.

Distribution: Frequent in hedges and thickets.

Therapeutic Usage:

Ethnopharmacology:

• Pulverized fruits are reportedly efficacious in treating jaundice and even in tuberculosis.
• Tribals apply root-bark paste on swollen testicles.
• Decoction of stem bark is prescribed in the treatment of cholera.
• They use crushed leaves against smallpox.

FAMILY COMBRETACEAE

Introduction

The combretum or bush willow family, Combretaceae, is a complex collection of flowering plants found worldwide, mostly in tropical and subtropical environments [7]. Around 20 genera and 600 species comprise this family, with several ecological adaptations and morphologies. Combretaceae occur in savannas, forests, and coastal areas [101]. Their opposing leaves and five-petal clustered flowers distinguish them. In many environments, these plants provide food and shelter for wildlife. Traditions use Combretaceae family members for medicinal purposes. Anti-inflammatory and ant-diabetic activities have been found in the African bush willow (*Combretum apiculatum*) [101]. Traditional medicine uses these herbs to treat various illnesses. Various societies value the Combretaceae species ethnobotanically. Some plants have culinary, dye, and ritual functions in addition to their medicinal importance [102]. Their strong roots in local traditions explain their vast range of relevance across areas and communities. Ethnoveterinary practitioners also have used Combretaceae plants [103]. Indigenous knowledge informs plant parts or extracts used for animal health. Wide distribution, diversified flora, and possible uses in medicine, cultural practices, and ecology make the Combretaceae family necessary in human culture and traditional knowledge systems.

• *Termination alata* Heyne ex Roth.

Local Name: Asana

Botanical Characteristics: Trees; leaves are alternate, simple, villous; flowers in spikes, small, yellowish brown; fruits drupe, glabrous, winged.

Distribution: Common in forest areas.

Therapeutic Usage:

Ethnopharmacology:

• Stem bark decoction is given twice a day to cure pneumonia.
• Stem bark decoction is also used as an antidote to snake venom. Fruits are also used to treat leprosy.

• *Terminalia arjuna* (Roxb, ex DC.) Wt. & Arn.

Local Name: Kahua

Botanical Characteristics: Trees; bark smooth, grey, flaking off; leaves alternate, simple, with two glands at the base, elliptic; flowers in axillary spikes; fruits drupes, ovoid, 2-5 winged; seeds stony solitary.

Distribution: Common in forests.

Therapeutic Usage:

Ethnopharmacology:

• Stem bark decoction is prescribed for restlessness and accelerated palpitation.
• Stem bark decoction twice a day is recommended for chronic fever.
• Ash of stem bark is frequently used as toothpowder to cure pyorrhea and other teeth problems.
• The juice of fresh leaves is used for earache.

• *Terminalia bellerica* (Gaertn.) Roxb.

Local Name: Baheda

Botanical Characteristics: Trees; the stem is hard, erect, and woody; leaves opposite, simple, entire, and exstipulate; flowers in the paniculate racemose inflorescence; fruits drupe, green; seeds non-endospermic.

Distribution: Found cultivated in the forest areas.

Therapeutic Usage:

Ethnopharmacology:

• Fruits are used in abdominal disorders and to check vomiting and loose motions.
• 10 gm of fruit powder is taken twice daily for three days to cure fever.
• Fruit powder is given for the treatment of cough in a dose of 1 gm with honey twice a day.

• *Terminalia chebula* Retz.

Local Name: Harra

Botanical Characteristics: Medium-sized deciduous trees with dark brown bark; leaves 10-18 cm long; flowers cream-coloured in terminal panicked cymes; drupe 3-5 cm long, ovoid or obovoid, ribbed.

Distribution: Cultivated in forest areas.

Therapeutic Usage:

Ethnopharmacology:

• Unripe fruits are used to control diarrhea and dysentery.
• 1 teaspoon of dried fruit powder is administered with warm water daily before bed to cure gastric and flatulence.
• Fruit paste with turmeric powder is applied in inflammation of the eyes.

FAMILY CUCURBITACEAE

Introduction

The gourd or cucumber family, Cucurbitaceae, is a complex collection of flowering plants worldwide, mostly in tropical and subtropical environments [104]. This family includes cucumbers, pumpkins, melons, and squashes, with 130 genera and over 900 species [105]. Cucurbitaceae plants are versatile and ecologically important, growing in arid deserts and rainforests. Their climbing or trailing vines, palmate leaves, and edible fruits distinguish them. Cucurbitaceae has had medical importance in traditional medicine [106]. Bitter melon (*Momordica charantia*) may help manage blood sugar [107]. These plants have helped traditional herbal treatments treat various illnesses. Many cultures value

Cucurbitaceae species ethnobotanically. Some plants provide unique-tasting fruits and vegetables in addition to medical purposes [108].

Additionally, certain Cucurbitaceae species have religious and ceremonial significance in different civilizations, indicating their profound ties to local customs. While not as abundant as other plant families, some Cucurbitaceae species may have been used in ethnoveterinary practices [109]. Indigenous knowledge typically informs animal health treatment with plant parts or extracts. The Cucurbitaceae family's global distribution, diversified flora, and usefulness in medicine, culinary arts, and culture make it essential in human culture and traditional knowledge systems.

• *Momordica charantia* L.

Local Name: Karela

Botanical Characteristics: Climbing annual herbs with simple slender tendrils; leaves cordate, orbicular; flowers unisexual, male flowers on 6 cm long peduncle, solitary, yellow; fruits 2.5–4 cm long, fusiform, tubercled.

Distribution: Cultivated in fields and kitchen gardens. It is also found occasionally in waste places near villages.

Therapeutic Usage:

Ethnopharmacology:

• Juice of raw fruit is taken orally in the morning to check diabetes and protect from heat strokes.
• Leaf extract mixed with the extract of *Azadirachta indica* (Neem) leaves is also used to cure diabetes.

FAMILY CUSCUTACEAE

Introduction

The Cuscutaceae, or Dodder family, is a rare family of parasitic flowering plants. Members of this small (100-170 species) group are parasitic and chlorophyll-deficient [110, 111]. The Cuscutaceae family can be found worldwide in tropical and temperate regions. Their threadlike, twining stalks cling to their hosts, sucking nutrition from them [112]. The parasitic nature and lack of chlorophyll in Cuscutaceae plants mean they are not typically used in medicine [113]. However, several cultures rely on types of dodders for their purported medicinal properties. Such applications occur far less frequently in parasitic plant families. The ethnobotanical significance of Cuscutaceae species in either culture or cuisine is negligible. Due to their parasitic nature, they are not cultivated or utilized in traditional culinary, ritual, or other practices. Because of the harm they cause to their hosts, members of the parasitic Cuscutaceae family are rarely used in ethnoveterinary practices [114]. The Cuscutaceae family is distinct in that its members are parasites that lack chlorophyll and instead obtain nourishment from their hosts [115].

• *Cuscuta reflexa* Roxb.

Local Name: Sarag baburi

Botanical Characteristics: Parasitic twines; stem glabrous, pale greenish yellow; leaves absent; flowers in racemiform cymes, solitary, white; fruits capsules; seeds orbicular black.

Distribution: Common on bushes.

Therapeutic Usage:

Ethnopharmacology:

• Tribals apply fresh paste on dislocated shoulder joints.
• The paste of the whole plant is applied to cure swelling of the testicles.
• Seed paste is taken orally as an anthelminthic.
• In case of jaundice, the plant is tied to the neck of the patient and is also spread on the bed. The vapours are inhaled.
• A warm paste of the plant is applied to the affected part.
• The paste of the whole plant is applied for relieving headaches.

FAMILY DIOSCOREACEAE

Introduction

Yams, or Dioscoreaceae, are a varied and economically important family of flowering plants. This family is essential in agriculture, traditional medicine, and culture, with 845 species worldwide [116]. Dioscoreaceae plants grow in tropical rainforests and temperate woodlands. Their heart-shaped leaves and climbing or twining growth behavior distinguish them. The most famous member of this family is the yam (*Dioscorea spp.*), a significant food crop worldwide [117]. Dioscoreaceae species have been used in traditional medicine. The yam may have anti-inflammatory and antioxidant qualities [118]. These plants have shown a potential to treat various illnesses. Dioscoreaceae plants are culturally significant in many communities. Besides their therapeutic applications, yams and other species are vital to traditional cuisine and culture [119]. Due to their deep local roots, these plants are prominent in rituals, ceremonies, and festivals. While not as common as other plant families, Dioscoreaceae species may have been used in ethnoveterinary practices [120]. Indigenous knowledge typically informs animal health treatment with plant parts or extracts [121]. The Dioscoreaceae family's global distribution, diversified flora, and relevance in agriculture, medicine, and culture make it essential in human culture and traditional knowledge [122]. As with any plant use, precise identification and toxicity knowledge are necessary for safe and effective use in many applications.

• *Dioscorea bulbifera* L.

Local Name: Belarakanda

Botanical Characteristics: Twining herbs; leaves are alternate, simple, ovate, cordate at the base, but always with an axillary bulbil; flowers in spikes; fruits are in capsules, quadrate oblong; seeds flattened with oblong wings above.

Distribution: Common in forest areas and hedges as vines.

Therapeutic Usage:

Ethnogynaecology:

• Paste of root tubers is prescribed 3 spoonfuls twice a day for about a week for abortion of pregnancy.

Ethnopharmacology:

• A powder of dried tubers is applied on sores.
• Powder of dried tubers is also administered to kill hair lice.
• Tribals apply a paste of tubers with ginger paste in a ratio of 3:2 externally on fractured bones.

Ethnoveterinary:

In Foot and Mouth cattle disease, the paste of tubers with a decoction of long peppers is effective.

FAMILY EUPHORBIACEAE

Introduction

The Euphorbiaceae family, or spurge family, is one of the largest and most diverse plant groups, with 300 genera and 7,500 species [123]. These plants are found worldwide, mostly in tropical and subtropical climates, and are highly adaptable. Euphorbiaceae plants grow in deserts, grasslands, rainforests, and coastal environments. They grow as herbs, shrubs, or trees and produce milky sap that can be hazardous [124]. Euphorbiaceae species are essential in traditional medicine. Castor bean (*Ricinus communis*) seeds and rubber tree latex have been utilized in traditional remedies [125]. These plants have helped traditional herbal treatments treat various illnesses. Euphorbiaceae plants are culturally significant in many societies [126]. Some species are utilized for cooking, crafts, and ceremonies in addition to medicine. Their relevance varies by location and community due to their deep roots in local traditions [126]. Euphorbiaceae species are also used in ethnoveterinary practices [127]. Indigenous knowledge typically informs animal health treatment with plant parts or extracts. The Euphorbiaceae family's global distribution, diversified flora, and potential medical, cultural, and ecological uses make it essential in human culture and traditional knowledge systems [69]. As with any plant use, precise identification and toxicity knowledge are necessary for safe and effective use in numerous contexts.

• *Emblica officinalis* Gaertn.

Local Name: Amla

Botanical Characteristics: Small or moderate-sized deciduous trees with greenish grey bark; leaves sessile subsessile, simple, elliptic or linear-oblong, mucronate, pale green; flowers greenish yellow in axillary fascicles on lower leaves; fruits globose, fleshy, green to pale yellow.

Distribution: Found occasionally along the edges of forests.

Therapeutic Usage:

Ethnopharmacology:

• Dried fruit powder is used for stomach troubles such as indigestion and constipation. Tribals use fruits in jaundice.
• One green fruit is taken twice daily for 2-3 days to treat scurvy.
• Dried fruit powder is given for diarrhea and dysentery.

• A fermented liquor prepared from fruit is used for jaundice, dyspepsia, and cough.
• A paste of fruit with honey is applied to the eyes for clarity of vision.

• *Jatropha curcus* L.

Local Name: Bakrenda

Botanical Characteristics: Shrubs; stems soft wooded with latex; leaves alternate, palmately lobed; flowers in cymose panicles; yellowish green; fruits capsule, triangular; seeds brownish black. Plants grow both from cuttings and from seeds.

Distribution: Common in villages, hedges, and roadsides, planted for fencing.

Therapeutic Usage:

Ethnopharmacology:

• Tribals use latex to cure eczema.
• Ethnic people use fresh latex with common salt (1:1) to fix loose teeth.
• Tender twigs are used for cleaning teeth.
• Seed oil is used for the treatment of tumors.
• To stop bleeding from cuts, fresh latex is commonly used.

• *Mallotus phillipensis* (Lamk.) Mueller Arg.

Local Names: Rohini, laali

Botanical Characteristics: Small trees; leaves simple, peltate, palmately veined; flowers in axillary racemes, unisexual; male flowers with many stamens, female flowers trilocular; fruits capsules with red glandular hairs; seeds smooth black.

Distribution: Found commonly mixed with *Shorea robusta* &*Terminalia alata* near the forests.

Therapeutic Usage:

Ethnopharmacology:

• Root bark paste is applied locally on rheumatic swellings.
• Fruits are used in destroying tapeworms.
• Dried fruit powder with long peppers (5:2) is given weekly to cure gall bladder stones.
• Fruits are administered for skin diseases.
• Red powder from fruits commonly known as Kamela is used with ghee to cure boils and blisters.

• *Phyllanthus fraternus* Webster in Contr.

Local Names: Jar amla, dalgola

Botanical Characteristics: Herbs; leaves alternate, pinnately compound; flowers unisexual, minute, green; fruits capsules from rows side underside leaves; seeds trigonous brown, longitudinally striated on the back side.

Distribution: Common as a weed in cultivated fields and forests.

Therapeutic Usage:

Ethnogynaecology:

Root paste is given to women with genito-urinary troubles.

Ethnopharmacology:

• Tribals use plant juice with curd to cure jaundice.
• Roots are given to sleepless children.
• Root decoction is prescribed to cure dysentery.

• *Putranjiva roxburghii* Wall. Tent.

Local Name: Pinpina

Botanical Characteristics: Moderate-sized evergreen tree, also known as child-life tree; bark grey; branches drooping; leaves alternate, obliquely elliptic-oblong or ovate, acute or shortly acuminate, glabrous; flowers unisexual, yellowish, male flowers subsessile, densely clustered; female flowers 1-3, long, pedicellate; drupes ellipsoid.

Distribution: Found occasionally in forest areas, in damp places.

Therapeutic Usage:

Ethnogynaecology:

Fruits are recommended to pregnant ladies to prevent abortion during pregnancy.

Ethnopharmacology:

• The infants and women use hard stoned fruits as charm and are noted for their supposed protection against harm and evil eyes.
• Extract of leaves is given twice a day to cure fever in children.

FAMILY FABACEAE

Introduction

The Fabaceae, or legume family, is one of the world's largest and most commercially important plant families. It is vital to agriculture, ecology, and traditional knowledge systems, with over 19,000 species worldwide [128]. Fabaceae plants grow in almost every habitat, from cold tundra to tropical rainforests [129]. Their complex leaves and gorgeous bilateral blooms distinguish them. This family includes herbs, shrubs, and trees, making it adaptable. Fabaceae species are essential in traditional medicine. Many members, like red clover (*Trifolium pratense*) and licorice (*Glycyrrhiza glabra*), have therapeutic benefits [130]. These plants have helped traditional herbal treatments treat various illnesses.

Fabaceae plants are culturally significant in many communities. They are essential in cooking protein-rich meals like beans, lentils, peanuts, and medicine [131]. Different societies may also value various Fabaceae plants for rituals and ceremonies, showing their profound ties to local customs. In ethnoveterinary practices, Fabaceae species are utilized. Due to their protein content, leguminous fodder crops are commonly fed to animals [132]. Indigenous knowledge governs the usage of plant parts or extracts for animal health. The Fabaceae family's global distribution, diversified flora, and vital functions in agriculture, medicine, culinary arts, and culture make it essential to human culture and traditional knowledge systems [133]. As with any plant use, precise identification and toxicity knowledge are necessary for safe and effective use in many applications.

- *Abrus precatorious* L.

Local Name: Gughuchi

Botanical Characteristics: Twining shrubs; stems slender; leaves pinnately compound; flowers in racemes, pink (in red seeded plants), white (in white seeded plants); fruits pod, flat; seeds polished.

Distribution: Common in thickets and hedges.

Therapeutic Usage:

Ethnogynaecology:

- Tribals give root paste of white seeded plants of about 3 gm with a paste of 7 long peppers to women twice daily as a cure for white discharge.
- Root extract, a root extract of *Asparagus racemosus* and *Cuscuta* is given orally

one teaspoon thrice a day for three days after menstruation to check conception.
• Seed paste when placed in cotton and inserted in the vagina causes abortion.

Ethnopharmacology

• Seed and root paste is applied externally on goiter goiter-affected area once a day until cure.
• One part of seed powder mixed with four parts of *Eclipta prostata* in mustard oil is given in dermatitis.

• *Butea monosperma* (Lamk.) Taub.

Local Name: Dhaka

Botanical Characteristics: Medium-sized trees; stem with irregular branching; leaves pinnately trifoliate, leaflets coriaceous, glabrous above, obovate, and pubescent; flowers in dense fascicles, racemose, orange colored; fruits pod, long; seeds oval, compressed, dark brown.

Distribution: Found in forest areas and grasslands.

Therapeutic Usage:

Ethnogynaecology:

They prescribe flower paste to pregnant women with acute diarrhea.

Ethnopharmacology:

• Tribals apply stem bark paste on the fractured bone.
• They apply the flower paste externally on the inflammation of the testicles.
• Rural persons give seed oil to children for intestinal worms.
• Tribals give root decoction with egg to children to treat night blindness.
• Seed paste is used for cooling effect. Dried flowers soaked in water are used for taking baths to prevent sunstroke.

FAMILY LAMIACEAE

Introduction

The Lamiaceae family, also known as the mint or deadnettle family, is a varied group of flowering plants with aromatic, culinary, medicinal, and decorative properties. With 236 genera and over 7,000 species, Lamiaceae are found worldwide, mostly in temperate and tropical climates [134, 135]. Lamiaceae plants grow in grasslands, forests, marshes, and alpine regions. They have square stems, opposite leaves, and glandular trichomes with fragrant oils [136]. The Lamiaceae family is essential in traditional medicine. Peppermint, basil, and oregano (*Mentha piperita*) are among its members with possible medicinal effects [137]. These plants have helped traditional herbal treatments treat stomach and respiratory disorders. Lamiaceae plants are culturally significant in many communities [138]. These herbs add flavor to foods globally in addition to their therapeutic uses. Due to their profound ties to local customs, Different societies may also value Lamiaceae plants for rituals and ceremonies. Indigenous knowledge is used to treat and prevent livestock and domestic animal illnesses with herbal remedies from these plants [139]. The Lamiaceae family's global range, fragrant diversity, and widespread use in medicine, culinary arts, and culture make it essential in human culture and traditional knowledge systems [140]. As with any plant use, precise identification and toxicity knowledge are necessary for safe and effective use in many applications. These plants belonging to the Lamiaceae family enhance our diverse array of medicinal herbs. Although these substances have a longstanding history of traditional usage for diverse health reasons, it is crucial to exercise prudence and seek advice from healthcare experts or herbalists regarding their safe and efficient application. Ongoing scientific research explores these plants, elucidating their active components and potential applications in health and wellness.

• *Salvia officinalis* **L.**

Local Name: Samudraphal

Botanical Characteristic: Perennial, evergreen under shrub with woody stems leaves grayish, flowers purple, root system well developed, flowers lavender blue, aromatic.

Distribution: Wild/Cultivated.

Therapeutic Usage:

Ethnopharmacology:

• It is an herb renowned for its characteristic fragrance and potential medicinal advantages. Sage has a long-standing history of use by different civilizations for its therapeutic benefits. It is widely acknowledged to possess antioxidant, anti-inflammatory, and antibacterial properties.

• It has been utilized to promote dental health, reduce throat discomfort, and regulate symptoms associated with menopause.

• Its leaves are utilized in infusions, teas, and as a culinary spice. Sage essential oil is commonly employed in aromatherapy due to its distinct earthy and vegetal fragrance. Although sage has potential health advantages, particularly in traditional medicine, additional scientific investigations are currently being conducted to comprehend its unique constituents and methods of action for different health uses.

• *Thymus vulgaris* L.

Local Name: Banajwain

Botanical Characteristic: It is a perennial subherb, 10-30 cm in height with slender, wiry and spreading branches. Leaves are small, evergreen, opposite, oblong- lanceolate, 5-10 mm long and 0.8-2.5 mm wide, highly aromatic. Flowers are purple or pink in color.

Distribution: Cultivated.

Therapeutic Usage:

Ethnopharmacology:

• It is a fragrant herb renowned for its applications in cooking and medicine. Thyme has a significant historical background in traditional medicine and is thought to have antibacterial, antioxidant, and anti-inflammatory characteristics.
• It is used to relieve respiratory ailments such as coughs and bronchitis, commonly found in cough syrups and herbal treatments. Thyme essential oil is highly valued for its ability to enhance immunological function and promote better blood flow. The herb's fragrant taste is highly valued when added to culinary preparations, while its essential oil is widely used in aromatherapy and massages for its ability to induce calm. Thyme, like other herbs, demonstrates its adaptability by its use in both traditional and modern contexts. Nevertheless, it is important to consult healthcare professionals or herbalists to guarantee the secure and efficient use of it in different health circumstances.

• *Ocimum basilicum* L.

Local Name: Dauna

Botanical Characteristics: Herbs; leaves opposite, simple, ovate, lanceolate, glandular; flowers in racemes, white or purple; fruits nutlets, black pit.

Distribution: Common in waste places under shades and in open places.

Therapeutic Usage:

Ethnopharmacology:

- To improve sexual weakness, tribals use root decoction with honey for a month.
- Stem decoction with a paste of long peppers is given to cure fever, cough & cold.
- Tribals put green leaf juice as drops in the nose and ear to cure nasal congestion and earache.
- Seed paste is applied on bees, wasps, and insect bites.

FAMILY LAURACEAE

Introduction

The Lauraceae family, or laurel or avocado family, is a varied and economically significant group of flowering plants. This family is essential in agriculture, forestry, and traditional knowledge systems, with 2850 species worldwide, mostly in tropical and subtropical climates [141, 142]. Lauraceae plants grow in rainforests, mountains, and coasts. Their evergreen, aromatic leaves and range of growth forms—trees, shrubs, and woody vines—distinguish them. Lauraceae species have medicinal value in traditional medicine. Bay laurel (*Laurus nobilis*) and cinnamon (*Cinnamomum verum*) are among the members used for their medicinal effects [143]. These plants have helped traditional herbal treatments treat digestive and inflammatory disorders [144]. Lauraceae plants are culturally significant in many communities. Bay leaves and cinnamon are popular spices and flavorings in addition to their medical purposes. Some Lauraceae species are also ritually and ceremonially important in different civilizations, indicating their profound ties to local traditions [145]. Lauraceae species may be used in ethnoveterinary practices. Indigenous knowledge typically informs animal health treatment with plant parts or extracts [121]. The Lauraceae family's global distribution, diversified flora, and widespread use in medicine, culinary arts, and culture make it essential in human culture and traditional knowledge systems [146]. As with any plant use, precise identification and toxicity knowledge are necessary for safe and effective use in numerous applications.

A notable constituent is *Cinnamomum verum*, commonly known as Ceylon Cinnamon, which is highly regarded for its fragrant bark and esteemed as a culinary seasoning. It has a profound historical background in traditional medicine, reputed to possess antioxidant, anti-inflammatory, and antibacterial characteristics. *Ceylon Cinnamon* promotes digestive health and regulates blood sugar levels, rendering it a favored option for ailments such as colds, flu, and gastrointestinal distress. The avocado tree, scientifically known as *Persea americana*, yields avocados that are abundant in nutrients and are widely recognized for their exceptional culinary and health advantages. These fruits are renowned for their high monounsaturated fats, vital minerals, and dietary fiber content. Avocados have benefits for cardiovascular health, skin health, and weight control. *Laurus nobilis*, also known as Bay Laurel, is a highly esteemed member of the Lauraceae family, renowned for its fragrant leaves commonly utilized in culinary applications. Bay leaves have possible therapeutic benefits, especially in aiding digestion, managing indigestion, and their culinary purposes. Traditional medicines have also been investigated for their potential to reduce inflammation and promote diuresis. *Persea gratissima*, commonly referred to as the Avocado Pear, is closely related to the avocado tree from a botanical perspective. The Avocado Pear's fruit is exceptionally nutritious and rich in monounsaturated fats, vitamins, and minerals. It is widely acknowledged for its capacity to enhance cardiovascular health, reduce cholesterol levels, and supply vital nutrients to sustain general health. The Lauraceae family encompasses a wide range of plant species that have been historically and now utilized in medicine and nutrition. Although many substances have been acknowledged for their medicinal characteristics in traditional medicine, current scientific investigations seek to examine better and authenticate their prospective health advantages. For the aim of using plants from the Lauraceae family for medicinal reasons, it is crucial to get advice from healthcare experts to guarantee the therapy's safety and effectiveness.

• *Litsea glutinosa* (Lour.) C. B. Robins

Local Name: Maida lakari, Kathbhilwa.

Botanical Characteristics: Evergreen trees; leaves alternate; flowers greenish yellow in umbellate heads.

Distribution: Frequent near cultivated fields.

Therapeutic Usage:

Ethnopharmacology:

• Seeds yield fat, which is used as an ointment for rheumatism.
• Paste of stem bark mixed with common salt is used externally to treat gout, inflammations, and sprains.
• Paste of stem bark with common salt is applied externally in bone fracture.

Ethnoveterinary:

- Gujjars externally use fresh stem bark paste. They apply this paste as plaster on the fractured bones of their cattle.

FAMILY LEEACEAE

Introduction

The Leeaceae family of flowering plants is small and less well-known than other plant families, yet they have unique characteristics and may be significant in traditional knowledge systems [147]. The 46 genera comprising the family Leeaceae are almost exclusively restricted to warmer regions [148]. The Leeaceae family is primarily distributed in tropical and subtropical woodlands. Some species also have fragrant blooms and leaves that alternate. Even though they are ecologically significant in their home regions, these plant families are not as widespread or well-documented as others. Leeaceae plants are rarely used in traditional medicine due to their rarity in their native environments [149]. Although it is less common than with other plant families, local communities may have relied on the healing properties of specific members of this family [150]. Leeaceae plants have substantial cultural meaning in their native environments but are not used as frequently in cuisine, rituals, or ceremonies as members of other plant families [151]. They have deep roots in the traditions of their respective regions. Rarely some members of the Leeaceae family play a role in traditional veterinary care for cattle and other animals [152]. Some plant families are better known and have wider traditional usage, whereas other species may only be of cultural or medical significance in their native range.

- *Leea indica* L.

Local Name: Chattri

Botanical Characteristics: A shrub or small tree leaves alternate, stalked with many leaflets, flowers greenish-white, fruits round, purple- black, stem with stilt roots, stipulate, leaflets ovate to lanceolate, cyme inflorescence.

Distribution: Wild.

Therapeutic Usage:

Ethnopharmacology:

- *Leea indica*, also referred to as Bandicoot Berry, is a plant that has a significant historical background in traditional medicine across several areas, notably in

India and Southeast Asia. The therapeutic benefits of this plant primarily originate from its leaves. They have historically been utilized for their inherent ability to reduce inflammation and alleviate pain.

- As part of conventional methods, the leaves are frequently pulverized and directly applied to the skin as poultices or pastes to relieve pain and inflammation linked to arthritis, joint discomfort, and traumas. The anti-inflammatory effects of Bandicoot Berry can be ascribed to the presence of bioactive chemicals present in its leaves. Although the scientific inquiry into these substances is still in progress, the plant's promise has been acknowledged by traditional wisdom for millennia. Bandicoot Berry is highly esteemed in indigenous medicinal practices due to its longstanding usage.
- The plant is frequently included in remedies, along with other traditional medicinal plants, to produce poultices or infusions for the treatment of localized pain and inflammation.

• *Gluta usitata* Wall.

Local Name: Amili

Botanical Characteristic: A medium to large tree, leaves dark green and forms a spreading crown, flowering occurs from December to March, fruits monolocular drupe, seeds germinate after 16-23 days of sowing.s

Distribution: Cultivated.

Therapeutic Usage:

Ethnopharmacology:

- It is a botanical species traditionally utilized in several Asian societies. The therapeutic qualities of this tree primarily stem from its bark. It is thought to have astringent properties, which are beneficial in managing ailments such as diarrhea and dysentery. The bark is frequently collected and processed into a decoction or infusion as part of conventional methods. The resultant liquid is subsequently ingested to alleviate gastrointestinal ailments.
- Its bark is said to possess astringent qualities that can aid in reducing diarrhea by constricting the intestines and minimizing the loss of fluids.
- The ancient utilization of its Bark to treat gastrointestinal ailments underscores the significance of specific plant species in offering organic therapies for prevalent health conditions. Although the prospective advantages of it have been acknowledged for many years, there is a scarcity of scientific investigation on the precise chemicals accountable for their medicinal impacts. Further research

is required to investigate the active chemicals and mechanisms contributing to its astringent characteristics. Before using Mango Bark-based treatments, it is recommended to seek guidance from healthcare professionals or traditional healers who possess expertise in producing and implementing such cures. Following this guide will guarantee the secure and efficient utilization of this herb for pharmaceutical intentions.

• *Leea guineensis* **G.Don.**

Local Name: Bahasa

Botanical Characteristics: Shrub or small tree leaves pinnately compound, leaflets are elliptical to lanceolate, flowers red berries and ovoid with 4-5 seeds, stem pithy, petioles 6-15 cm long, flowers cyme and red.

Distribution: Wild.

Therapeutic Usage:

Ethnopharmacology:

- This herb has historically been used for its putative anti-inflammatory effects. Leaves and other plant components are frequently employed in traditional medicines, such as poultices and infusions, to mitigate illnesses linked to pain and inflammation.
- Leaves are traditionally crushed or macerated to produce pastes or infusions. Subsequently, these substances are administered externally to the specific region to alleviate symptoms associated with ailments such as arthritis, muscular discomfort, or dermal inflammation. The purported anti-inflammatory properties of it are found in its stem due to the presence of bioactive chemicals in the plant.
- The traditional utilization of this plant in pain and inflammation management highlights the significance of indigenous plant species in traditional medicine. Although the efficacy of these conventional treatments has been acknowledged for many years, additional scientific investigation is required to clarify the precise chemicals accountable for their therapeutic capabilities and to authenticate its applications. When using African Leea as a traditional cure, it is crucial to be cautious and consult healthcare professionals or traditional healers who are well-informed about its traditional usage. This will help guarantee that the treatment is both safe and successful.

• *Leea sambucina* **(L.) Willd.**

Local Name: Berry

Botanical Characteristics: A shrub or small tree, up to 15 mt tall, leaves alternate, stalked, 2-3 pinnate, flowers greensh white, fruts round, purple black, stem with stilt roots, inflorescence corymbose cyme, calyx green.

Distribution: Cultivated.

Therapeutic Usage:

Ethnopharmacology:

- *Leea sambucina*, is a plant with a history of traditional use in various cultures. This plant is believed to possess diuretic properties, and extracts derived from it are used in traditional remedies to promote urination and address conditions related to fluid retention. In traditional practices, different plant parts, such as the leaves or roots, are often used to prepare decoctions or infusions. These preparations are consumed to increase urine production, which is believed to help relieve conditions like edema (swelling due to fluid retention).
- The traditional use of it as a diuretic highlights the diverse array of plants with potential therapeutic properties in traditional medicine. While traditional knowledge recognizes its diuretic effects, the specific compounds responsible for this property and the mechanisms of action are areas that require further scientific investigation. As with all traditional remedies, it is recommended to consult with healthcare professionals or traditional healers knowledgeable about its traditional applications to ensure the safe and effective use of the plant for medicinal purposes. Further research may better understand the plant's bioactive compounds and potential health benefits.

• *Leea asiatica* (L.) Ridsdate

Local Name: Golkandara

Botanical Characteristics: Shrubs; leaves pinnately compound, turn red before drying in December; flowers in corymbose cymes; greenish white; fruits berries, blue-black when ripe; seeds stony.

Distribution: Common in forests along forest roads.

Therapeutic Usage:

Ethnopharmacology:

- To cure pneumonia, root bark paste with capsicum is applied to the chest.
- Tribals prescribe leaf decoction with the paste of black pepper for a week for urinary problems.

• Green leaf juice is applied locally on ringworm-affected areas.

Ethnoveterinary:

• Ethnic people apply root paste on the septic wounds of their cattle.

• *Leea macrophylla* Roxb.

Local Name: Lathigaja

Botanical Characteristics: Robust shrubs; leaves simple, ovate; flowers in terminal corymbs, whitish green; fruits berries, black when ripe; seeds stony.

Distribution: Common in forest areas

Therapeutic Usage:

Ethnogynaecology:

• Root powder is given to lactating mothers for better flow of the mother's milk.

Ethnopharmacology:

• Dried root powder is given to cure diarrhea and dysentery.
• Root paste, salt and goat milk are used as a bone fracture bandage.

Ethnoveterinary:

• Tribals apply leaf paste on the fractured bones of their pet animals.

FAMILY LECYTHIDACEAE

Introduction

The Lecythidaceae family, known as the Brazil nut or monkey pot family, is a tiny flowering plant family with 28 genera and around 300 species [153]. These plants are tropical, especially in the Americas and Southeast Asia, and have unusual fruit capsules. Tropical rainforests and other humid habitats are home to Lecythidaceae plants [154]. Their vast, beautiful blooms, frequently with many petals and woody fruit capsules with seeds distinguish them. Its edible seeds make the Brazil nut tree (Bertholletia excelsa) one of the most famous members of this family [155]. Although less prevalent than other plant families, local communities may have used some species of this family for their therapeutic benefits. Lecythidaceae plants may have cultural value in their natural regions, especially in communities where they are used regularly [154]. For instance, the Brazil nut is gathered for its nutritious seeds, utilized in cooking, and traded abroad. Some Lecythidaceae plants are ritual or ceremonial, reflecting their local traditions [156]. Lecythidaceae species are rarely used in ethnoveterinary practices for cattle and animal care [157]. Overall, the Lecythidaceae family is a small, unusual tropical plant family. Some species, like the Brazil nut tree, are economically and culturally significant, but others lack the global popularity or broad traditional applications of other plant families [158]. Here are some medicinal plants from the Lecythidaceae family, along with additional information:

• *Gustavia augusta* L.

Local Name: Kumbhi

Botanical Characteristics: Medium

Distribution: Cultivated.

Therapeutic Usage:

Ethnopharmacology:

• It is a tropical tree and bark, and leaves have been used for medicinal purposes. Nevertheless, the precise medical uses and the exact active substances accountable for therapeutic benefits remain ongoing research topics.
• It has traditionally been utilized to produce remedies to address various health conditions. Extracts from this plant are thought to have qualities that may be anti-inflammatory or analgesic, which could be beneficial for illnesses related to pain and inflammation. The safety and effectiveness of the products have not been thoroughly examined or confirmed in controlled clinical research. Before contemplating the utilization for medicinal intentions, seeking guidance from healthcare professionals or traditional healers who possess expertise in the plant's traditional uses is recommended.

• *Barringtonia acutangula* (L.) Gaertn.

Local Name: Sunaffar

Botanical Characteristics: Small glabrous trees; leaves 12-16 x 4.5 cm, short petiolate, obovate- oblong, glabrous; flowers dark pink in 15-20 cm long slender racemes; fruits 2-3 cm long, quadrangular, crowned by persistent calyx.

Distribution: Frequent in swampy areas and near river banks.

Therapeutic Usage:

Ethnopharmacology:

• Kernels are made into powder and 5 gm is taken with butter twice a day for three days in the treatment of diarrhoea.
• Seeds are pulverized and used as snuff for relief from headaches.
• Leaf paste is applied on the forehead for the same.
• Twig is chewed for toothache.

FAMILY LILIACEAE

Introduction

Liliaceae, the lily family, is a complex group of flowering plants recognized for its decorative beauty and historical use in medicine and ethnobotany. This family has 290 genera and nearly 4,000 species in temperate environments worldwide [159, 160]. Liliaceae plants grow in grasslands, forests, alpine meadows, and deserts. They have parallel-veined leaves and gorgeous six-petalled blooms. Liliaceae has traditionally been used in medicine, such as Madonna lily (*Lilium candidum*) [161, 162]. These plants have helped traditional herbal treatments treat digestive and skin disorders. Some species are grown as ornamentals in gardens and floral arrangements in addition to their medical applications. Due to their profound ties to local customs, Liliaceae plants are ritually and ceremonially important in several civilizations [145]. Liliaceae species may have been used in ethnoveterinary practices however not as often as other plant families. Indigenous knowledge typically informs animal health treatment with plant parts or extracts [163]. The Liliaceae family's global distribution, diversified flora, and possible uses in medicine, horticulture, and culture make it essential in human culture and traditional knowledge systems [164]. As with any plant use, precise identification and toxicity knowledge are necessary for safe and effective use in numerous contexts. The Liliaceae family, also called the lily family, encompasses a diverse range of plants, several of which have been traditionally utilized for their therapeutic attributes. The plants belonging to the Liliaceae family have served diverse functions in traditional and herbal medicine. Although certain plants such as aloe vera and garlic are well acknowledged for their beneficial effects on health, it is important to use caution when dealing with Sea Squill, as it contains harmful compounds. Before utilizing any botanical medicines or supplements for therapeutic intentions, seeking guidance from healthcare specialists or herbalists is imperative. Below are a few therapeutic plants belonging to the Liliaceae:

• *Aloe barbadensis* **Mill.**

Local Name: Gwar ka phatta

Therapeutic Usage:

Ethnopharmacology:

• Aloe vera is a renowned succulent plant characterized by gel-loaded leaves. It has been utilized for decades to alleviate skin irritations, burns, and wounds owing to its anti-inflammatory and moisturizing characteristics.
• In addition, aloe vera is used in several traditional herbal medicines for its potential internal health advantages, such as assisting digestion and enhancing bowel regularity. This plant is extensively grown and recognized for its various uses in medicine and cosmetics.

• *Polygonatum multiflorum* **(L.) All.**

Local Name: Satabari

Botanical Characteristics: A perennial herb leaves alternate, flowers white, bell shaped, underground rhizome is present.

Distribution: Wild/ Cultivated.

Therapeutic Usage:

Ethnopharmacology:

• It is an herb that has historically been employed for its purported anti-inflammatory effects.
• Its rhizome is frequently used as a natural treatment for joint and muscle pain.

• *Urginea maritima* **(L) Baker**

Local Name: Samudri Pyaaz

Botanical Characteristic: Herbaceous perennial, succulent plant- one of the largest known bulbs of 25 cm diameter, the inflorescence is long, narrow, and lobed with small flowers, white or pink or red in color.

Distribution: Wild

Therapeutic Usage:

Ethnopharmacology:

- It has traditionally been used in medicine for its potential cardiac and expectorant properties. Nevertheless, it is crucial to acknowledge that it also harbors poisonous substances and should solely be employed under the supervision of experts.

• *Allium cepa* L.

Local Name: Pyaz

Botanical Characteristics: Herbs; bulbs thick, globular, tunics membranous, space tall, hollow and leafy near the base; leaves in two rows, shorter than the scape; umbel globular, many-flowered; pedicels 4-5 times as long as the flowers; perianth white; segments ovate-oblong, acutes filaments longer than the perianth.

Distribution: Cultivated in fields and kitchen gardens.

Therapeutic Usage:

Ethnopharmacology:

- Fresh juice of bulb mixed with lemon juice (1:1) and is taken orally for indigestion, nausea and vomiting.
- Fresh juice of bulb mixed with salt is used in colic and scurvy.
- Fresh juice is also effective in piles.
- The bulb is cooked in vinegar and is given in jaundice, spleen enlargement and dyspepsia.
- Extract of bulbs and fresh herb is used against flatulence and dysentery.

• *Allium sativum* L.

Local Name: Lassun

Botanical Characteristics: Herbs; leaves flat, scape slender; spates long beaked; heads bearing bulbils and flowers; sepals lanceolate, acuminate, inner filaments 2 toothed.

Distribution: Cultivated in fields and kitchen gardens.

Therapeutic Usage:

Ethnopharmacology:

- Cloves are swallowed empty stomach early in the morning for arthritis.
- Cloves dipped in vinegar are eaten for proper digestion and appetite.
- 1-2 fresh cloves and honey are swallowed in the morning to check aging and enhance vitality.
- 1-2 bulb lets are crushed and administered orally to expel intestinal worms.
- Garlic is fried in mustard oil and the oil is applied in case of rheumatic pains.

• *Asparagus racemosus* Willd.

Local Name: Satavar

Botanical Characteristics: Woody, scan dent, much branched spinous shrub climbers, with tuberous roots; leaves cladodes in tufts of 2-6; flowers white, racemes; fruits berries, globose.

Distribution: Found frequently in the forests, straggling on small trees.

Therapeutic Usage:

Ethnogynaecology:

- Tender shoots and tuberous roots are taken as vegetables and prescribed to increase lactation in mothers with newly borne.

Ethnopharmacology:

- About 10 gm of root powder and cow's butter are given as antacids twice a day and used as a remedy for ulcers.
- Root powder mixed with *Acorus calamus*, *Allium sativum, Centella* spp. along with honey is given to treat unconsciousness and epileptic attacks.

Ethnoveterinary:

- Roots are given as vulnerary for diarrhoea and dysentery in animals.

• *Urginia indica* (Roxb.) Kunth

Local Name: Patalu, janglikanda

Botanical Characteristics: Herbs, perennial; adventitious fibrous roots; stem underground, tunicated bud; leaves radical, numerous, sword shaped; monochasial cymes, arranged in umbellate fashion; flowers small, white, bisexual; fruit membranous capsule.

Distribution: Common is forest areas.

Therapeutic Usage:

Ethnogynaecology:

• Bulb paste is prepared with sugar candy and one spoonful is taken once a day for a week to treat leucorrhea.
• Corm paste is put into cotton and inserted in the vagina to cause abortion.

Ethnopharmacology:

• Half of the bulb is ground with 10 black peppers in 50 gm of pure-ghee. This is given in three doses daily as an antidote for snake venom.
• The bulb is also used in bronchial troubles.

FAMILY LOGANIACEAE

Introduction

The Loganiaceae family, with 13 genera and 400 species, includes ecological, medicinal, and ethnobotanical flowering plants [165]. Tropical and subtropical environments are home to these plants. Loganiaceae flourish in forests, savannas, and wetlands. They have opposing leaves and tubular flowers that attract pollinators with brilliant colors. Some Loganiaceae species have been utilized in traditional medicine, mostly in their native regions. For instance, the Strychnine tree (*Strychnos nux-vomica*) has been tried for its potential therapeutic benefits but is toxic [166, 167]. In their native locations, Loganiaceae plants are culturally significant; they are not as crucial in culinary traditions, rituals, or ceremonies as other plant families. They are generally tied to local communities and customs. In ethnoveterinary practices, Loganiaceae species are rarely used in cattle and animal care [168]. The Loganiaceae family is a small, poorly investigated tropical and subtropical plant family. Some plant families are more prominent and have extensive traditional usage than others, yet some species have limited medicinal or cultural value in their native areas.

• *Strychnos nux-vomica* L.

Local Name: Kuchla

Botanical Characteristics: Trees; branches with axillary spines; leaves are opposite, simple, broadly elliptic; flowers in compound cymes, greenish; fruits berries, globose, shining orange-red when ripe; seeds many, compressed, one side convex, covered with silky hairs.

Distribution: Not very common in forests and village sides.

Therapeutic Usage:

Ethnopharmacology:

- Root bark paste or seed paste is applied on the snake-bitten parts. 1 spoonful of extract is given thrice a day orally as an antidote for snake poison.
- Seed decoction with honey is given to cure night wetting in children.
- The leaves are applied as poultices on sloughing wounds and maggot-infested ulcers.

Ethnoveterinary:

- Dried seed powder is given to horses as a tonic.
- Root and bark when fed to cows, impart a bitter taste to the milk, which is credited with digestive and tonic properties.

FAMILY LYGODIACEAE

Introduction

The Lygodiaceae family, or climbing ferns, has specialized adaptations for climbing and scrambling in many habitats [169]. Lygodiaceae ferns are found worldwide, especially in tropical and subtropical climates. Their slender, twining fronds and ability to climb and adhere to trees and rocks distinguish them. This adaption lets them grow in forests, wetlands, riverbanks, and disturbed places. Lygodiaceae ferns are rarely used in traditional medicine [170]. Due to their lack of well-documented medicinal characteristics, these ferns are rarely used in herbal treatments or traditional therapy [28]. Lygodiaceae plants are not culturally or culinary crucial in most countries. They play a lesser role in rituals, rites, and daily life than other plant families. Lygodiaceae species are rarely used in ethnoveterinary practices and are not typically connected with cattle and animal care [171]. In conclusion, the Lygodiaceae family of climbing ferns is unique. Due to their adaptability, they contribute to ecosystems, but human culture and traditional knowledge systems do not utilize them medicinally, ethnobotanically, or ethnoveterinary.

Lygodium japonicum, or Japanese climbing fern, is an example of such a plant. The utilization of this fern has been a longstanding practice in Chinese medicine. The rhizomes and fronds of this plant are thought to possess diuretic and antipyretic effects. Herbal infusions and decoctions mitigate symptoms of ailments such as fever and edema. Moreover, it is believed to be beneficial in treating urinary tract infections. It is important to acknowledge that the therapeutic application of the Lygodiaceae family plants is somewhat restricted compared to more widely recognized families of medicinal plants.

Furthermore, rigorous scientific research may not firmly establish the effectiveness and safety of certain herbs for medical use. Therefore, it is recommended to exercise caution and seek advice from healthcare specialists before using plants from this family for medical purposes. Although Lygodiaceae may not hold the same significance level in traditional medicine as many other plant families, it is a fascinating collection of ferns with distinctive traits, rendering it a captivating subject of botanical research and intrigue.

• *Lygodium flexuosum* (Linn.) Swartz.

Local Name: Bartant

Botanical Characteristics: Scan dent ferns with creeping rhizomes; leaves pinnately compound sparsely hairy, veins forked; sporangia spicately biseriate.

Distribution: Widespread in dense forest areas.

Therapeutic Usage:

Ethnopharmacology:

- Roots are dried and powdered; a pinch of powder is given once daily for 15-20 days to cure anaemia.
- 15 gm of root paste is mixed in one glass of water and given twice a day to cure diarrhoea.
- Root paste is taken in the treatment of stomach pain.
- Tribals apply fresh leaf paste on burn wounds.

FAMILY LYTHRACEAE

Introduction

The Lythraceae or loosestrife family has about 600 flowering plant species [172]. This family has herbaceous and woody species and is distributed worldwide, especially in tropical and temperate climates. Lythraceae plants are widespread, from riparian zones and forests to grasslands and meadows [173]. They have opposite or whorled leaves and gorgeous four- to six-petal flowers, making them easily distinguishable. Lythraceae plants have been utilized medicinally for ages, mostly in untamed settings. The purple loosestrife (*Lythrum salicaria*) may be therapeutic as an astringent and anti-inflammatory [174]. There is less evidence that they are medicinal than other plant groups. Lythraceae plants may have cultural value in their home countries but are rarely employed in Western cuisine, religion, or ceremony [175]. They are more significant because they reflect regional or local communities and customs. The ethnoveterinary usage of Lythraceae species is unknown, and these plants are not frequently used in traditional cow and animal care [176]. The Lythraceae family has moderate variety and is dispersed worldwide. Several species have cultural or medicinal value in their native areas, but their benefits are limited compared to other plant families. Remember that even within a family, qualities and practical applications might vary.

• *Lawsonia inermis* L

Local Name: Mehandi

Botanical Characteristics: Spinecent shrubs; tap root system; stems branched, herbaceous or woody; leaves small, oval, lanceolate; flowers small, dingy, strongly fragrant.

Distribution: Cultivated in homes as a hedge plant, also found in wild states.

Therapeutic Usage:

Ethnopharmacology:

- Leaf decoction is effective as gargles for sore throat.
- Bark decoction is given to patients who have jaundice, enlargement of the spleen, and some types of skin diseases.
- The flower leaf and young shoot extract are helpful in leprosy.
- In dysentery, leaf juice, along with black pepper and curd is given for two days.

FAMILY MALVACEAE

Introduction

The Malvaceae family, or mallow family, is a diverse and ecologically significant collection of flowering plants with economic, medical, and cultural relevance. The Malvaceae family, with 244 genera and 2440 species, is found worldwide in tropical and temperate climates [177, 178]. Malvaceae thrive in deserts, grasslands, marshes, and rainforests. They have palmate or lobed leaves and spectacular five-petal blooms. Hibiscus, okra, and cotton are well-known members of this family [179]. Traditional medicine historically and currently uses Malvaceae. Hibiscus, marshmallow, and other members have been employed for their therapeutic benefits. Traditional herbal treatments use these plants to treat respiratory and digestive issues. Malvaceae plants are culturally significant in many communities. Besides their medical benefits, several species, like okra pods, are grown for their edible components and are staple foods in many cuisines [180, 181]. Malvaceae plants may also be ritually and ceremonially important in different societies, demonstrating their profound ties to local traditions [145]. Malvaceae plants may be used in ethnoveterinary practices. Indigenous knowledge typically informs animal health treatment with plant parts or extracts [163]. The Malvaceae family's global distribution, diversified flora, and widespread use in medicine, gastronomy, and culture make it essential in human culture and traditional knowledge.

• *Abutilon indicum* (Linn) Sweet

Local Names: Kanghi, Pilibuti

Botanical Characteristics: Shrubs; stem pubescent; leaves alternate, simple palmately lobed, ovate, pubescent; flowers axillary, solitary, yellow; fruits capsules, hairy; seeds reniform, black.

Distribution: Widespread in waste places, roadsides, under sheds, or exposed places.

Therapeutic Usage:

Ethnopharmacology:

- Tribals apply the seed oil as a cure for scabies.
- They take dried seed powder in warm water (3:2) as a laxative.
- Leaf paste is applied to the abscess till cured.
- Leaf paste is applied externally on the itching part especially during the night.

Ethnoveterinary:

• Leaf paste is applied externally to kill lice in pet animals.

• *Sida rhombifolia* L.

Local Name: Lal bariara

Botanical Characteristics: Erect herbs; leaves are alternate, simple rhomboid-obovate; slightly hairy above; flowers in axillary racemes, yellow or white; fruits capsules; seeds smooth, black.

Distribution: Common as wild in waste places.

Therapeutic Usage:

Ethnopharmacology:

• Root bark paste is prescribed to treat rheumatism.
• Root paste is used as an antidote to snake venom.
• Fresh leaf juice is administered as Dettol to stop bleeding from new cuts.

FAMILY MELIACEAE

Introduction

The flowering plant Meliaceae family, or the mahogany or neem family, is varied and economically significant. With 53 genera and approximately 1400 species, Meliaceae members are found in tropical and subtropical regions and are important in medicine, forestry, and traditional knowledge [182, 183]. Meliaceae plants grow in rainforests, dry woodlands, coastal environments, and savannas. Their pinnate foliage and fragrant flowers distinguish them. Mahogany, neem, and chinaberry are well-known members of this family. The Meliaceae family is essential in traditional medicine. Many members, including Neem, have therapeutic benefits [184, 185]. Traditional herbal medicines use these plants to treat skin diseases, digestive troubles, and pest control due to their insect-repelling characteristics. Meliaceae plants are culturally significant in many communities. Besides its medical properties, several species like mahogany are valued for their furniture and timber construction [186]. Different societies may also value various Meliaceae species for rituals and ceremonies, showing their profound ties to local customs. Meliaceae species are also used in ethnoveterinary practices. Indigenous knowledge typically informs animal health treatment with plant parts or extracts [187]. The Meliaceae family's global distribution, diversified flora, and widespread use in medicine, forestry, and culture make it essential in human culture and traditional knowledge systems [188]. As with any plant use, precise identification and toxicity knowledge are necessary for safe and effective use in many applications.

• *Azadirachta indica* A. Juss.

Local Name: Neeba

Botanical Characteristics: Large tree; leaves alternate, pinnately compound; flowers in axillary panicles, white; fruits drupes; seeds pendulous.

Distribution: Common in forests, roadsides, and open places.

Therapeutic Usage:

Ethnopharmacology:

• For the treatment of jaundice, one spoonful of roasted flowers is taken along with sugar thrice a day for about five days.
• Leaf paste is applied locally on the chickenpox-affected part of the body.

• For the anthelmintic cure, leaf extract and oil of seeds are taken orally on an empty stomach in the morning.
• Leaf juice and root decoction are taken orally thrice daily to cure fever.
• Leaf paste and oil is given orally for one month to control diabetes.

Ethnoveterinary:

• Green leaf paste is applied externally to kill parasitic insects on pet animals' bodies.

FAMILY MENISPERMACEAE

Introduction

The Menispermaceae, an exciting plant family, has many species worldwide. Due to its rich botanical traits and distinct ecological contributions, this family has received attention in traditional medicine, and ethnoveterinary practices [189, 190]. Menispermaceae plants thrive in tropical and subtropical regions on multiple continents, demonstrating their flexibility. Vine or woody climbers, these plants weave through rainforests and other environments. Its global relevance comes from its distribution throughout the Americas, Asia, Africa, and Australia [191]. Menispermaceae's medicinal and ethnoveterinary uses are fascinating. Indigenous societies have used many species in this family for generations as medicine [192]. These plants include alkaloids and other bioactive chemicals that have medicinal potential. Traditional therapy has used Menispermaceae plants to cure fevers, digestive issues, and parasite diseases in humans and animals. Menispermaceae species have helped livestock and domestic animals in ethnoveterinary medicine [193, 194]. The use of these herbs to heal animal illnesses shows their cultural and practical value in rural communities. The ecological services of Menispermaceae are notable. Climbers and woody vines provide structural diversity in forest canopies, sheltering biodiversity [195].

Nitrogen-fixing Menispermaceae plants improve soil fertility and ecosystem health [196]. Their significance in pollinator recruitment and seed dissemination strengthens their habitat relevance. In conclusion, the Menispermaceae family is fascinating and has a long medicinal and ethnoveterinary history. These plants have various roles in ecosystems and human cultures worldwide due to their extensive range, distinct ecological contributions, and cultural value.

• ***Cissampelos pareira*** L.

Local Name: Madraichi

Botanical Characteristics: Perennial climbers, with hairy branches; leaves ovate-orbicular; inflorescence pendulous, up to 6 cm long; flowers greenish yellow; fruits drupes, obovoid.

Distribution: Common near the forests on wastelands and on roadsides.

Therapeutic Usage:

Ethnogynaecology:

To subsidize labor pain, 3 spoonfuls of plant decoction is taken orally while shoot paste is applied externally on the abdomen of pregnant women.

Ethnopharmacology:

• Leaves and roots are used as an antidote for snakebite and scorpion stings.
• Root decoction is taken thrice a day for three days in intestinal worms.
• Dried root powder in 1 teaspoonful is given with water for 15 days to cure peptic ulcers.
• The powdered root is given twice a day to treat chronic asthma.
• Aqueous root extract is given thrice a day for three days for fever.

• *Tinospora cordifolia* (Willd.) Miers

Local Name: Gurch

Botanical Characteristics: Perennial herbaceous vine, leaves simple, alternate, inflorescence pendulous, flowers unisexual, fruits ovoid.

Distribution: Widely distributed on trees anywhere.

Therapeutic Usage:

Ethnopharmacology:

• The dried stem is very frequently consumed as decoction along with ginger and cloves by patients claiming fever, cold, and cough and to improve immunity.
• Tribal people sometimes take tea made from leaves to recover from common flu.
• Its decoction is very much useful to lower blood sugar levels.

FAMILY MIMOSACEAE

Introduction

The Mimosaceae, a diverse and ecologically significant plant family, has fascinated botanists, ethnobotanists, and traditional healers worldwide [197]. This family is important in traditional medicine and ethnoveterinary practices because of its wide distribution and adaptability. The Mimosaceae family thrives in environments throughout continents [198]. From little shrubs to tall trees, these plants thrive in many climates. They thrive in desert Africa and Australia and lush rainforests in South America and Asia, demonstrating their adaptability. The medicinal and ethnoveterinary relevance of the Mimosaceae family is intriguing. Alkaloids and tannins in many plants in this family make them useful in traditional medicine [199]. Indigenous peoples have used these herbs for generations to treat many human and animal illnesses. Mimosaceae plants help people and animals worldwide in wound healing, pain alleviation, gastrointestinal disorders, and parasitic infections [200]. These herbs treat cattle ailments in ethnoveterinary medicine, improving rural populations' health and livelihoods. Mimosaceae species are culturally and practically crucial because generations carry down animal care knowledge [201]. The Mimosaceae family also provides essential ecological functions. Some species fix nitrogen, improving soil fertility and plant neighbors [202]. They also feed and shelter insects and birds, boosting biodiversity and ecosystem function. The Mimosaceae family is a botanical gem due to its wide range, diverse ecological contributions, and significant effects on traditional medicine and ethnoveterinary practices [203]. Researchers and local populations are fascinated by the family's tenacity and adaptability, which allows it to flourish in varied circumstances.

• *Acacia catechu* (Linn. f.) Willd.

Local Name: Khadira

Botanical Characteristics: Moderate-sized trees with rough dark colored bark and hooked paired spines; pinnae 15-20 pairs or more; pinnules numerous, linear-oblong; flowers white, fragrant in cylindrical axillary spikes; pods 5-5.5 cm long, flat, acute dark brown.

Distribution: Found commonly in the area along with Sheesham trees.

Therapeutic Usage:

Ethnopharmacology:

• Gum is extracted from the bark and is applied to the wounds.
• Leaf paste is mixed with castor oil and is applied locally on the boils or burns.
• Powdered catechu is mixed with mustard oil and used as an ointment for obstinate ulcers and leprous affections.

• *Albizzia lebbeck* (L.) Benth

Local Name: Sarin

Botanical Characteristics: Large deciduous trees with spreading crown; leaf rachis 12-15 cm long, pinnae 2-6 pairs, 10-15 cm long, leaflets 6-9 pairs; flowers white in corymbose racemes; pods 20-25 cm long, flat, thin, firm, 6-10 seeded.

Distribution: Found commonly on roadsides as avenue trees.

Therapeutic Usage:

Ethnopharmacology:

• Tribals apply stem bark decoction as a cure for septic.
• They apply dried bark powder to kill lice.
• Seeds are used in piles and seed oil is used for leprosy.
• Flowers are recommended for boils, swellings, and antidote for poison.

Ethnoveterinary:

• Tribals apply the decoction of the stem bark on cattle sore.

• *Pithecellobium dulce* (Roxb.) Benth. in Hook.

Local Name: Jungal jalebi

Botanical Characteristics: Spiny trees; leaves pinnately compound; flowers in heads, white; fruits pod, twisted, reddish when ripe; seeds compressed arillate.

Distribution: Common in hedges.

Therapeutic Usage:

Ethnogynaecology:

Stem bark decoction with black pepper paste (5:3) is given to treat leucorrhea in women

Ethnopharmacology:

• Tribals prescribe root bark decoction as a cure for dysentery.
• Tribals give ripened fruits to their children for intestinal worms.
• Fresh leaf paste is applied on septic wounds and boils.

FAMILY MORACEAE

Introduction

The vast and commercially important Moraceae family is present worldwide and affects many human and environmental interactions. This family has many species and is found across continents and environments. Moraceae are found in Asia, Africa, the Americas, and Oceania, demonstrating their adaptability [204-206]. The Moraceae family is essential in medicine, ethnobotany, and ethnoveterinary practices [207]. Many species in this family have been used in traditional medicine to cure human diseases for ages. They also provide food, fiber, and colors, essential to indigenous ethnobotanical knowledge. Ethnoveterinary practices use these plants to treat and control animal health [208, 228-231]. In addition to their cultural value, Moraceae plants provide ecological services. Some species fix nitrogen, improving soil fertility and supporting nearby vegetation. They also support local biodiversity by providing habitat and food for insects, birds, and animals [209]. Finally, the Moraceae family is diversified and adaptive, with a global presence. Their presence on multiple continents and use in traditional medicine, ethnobotany, and ethnoveterinary care demonstrate their value to human communities. Their ecological contributions emphasize their importance in biodiversity and ecosystem health in their regions.

• *Ficus racemosa* L.

Local Name: Gulariya

Botanical Characteristics: Deciduous medium-sized trees; leaves alternate, simple, elliptic-lanceolate; flowers in receptacles, shortly pedunculate, pyriform, male, female, and gall flowers together in one receptacle; fruits hypanthodium, reddish when ripe.

Distribution: Common as planted and self-sown trees on roadsides.

Therapeutic Usage:

Ethnogynaecology:

• Decoction of root bark is given to lactating mothers to promote secretion and flow of breast milk.
• Cold infusion of leaves and ripe fruits is given for 7 days to cure leucorrhea.

Ethnopharmacology:

• Tribals give stem bark decoction along with long pepper paste in the treatment of kidney stones.
• Fresh juice of leaves is given with water for about 10 days to treat diabetes.
• In dysentery, the latex of a plant with sugar is given.

FAMILY NYCTAGINACEAE

Introduction

The distinctive and diversified Nyctaginaceae family of flowering plants provides remarkable insights into botany, ethnobotany, and ecological functions [210]. Nyctaginaceae plants on multiple continents are known for their ecological benefits. The Nyctaginaceae family thrives in arid deserts and tropical rainforests worldwide. Their exceptional adaptability shows these plants' ecological versatility to varied environmental circumstances. They thrive in North and South America, Africa, and portions of Asia due to their adaptability to many temperatures and geographies [211, 212]. Nyctaginaceae plants have influenced ethnobotany and traditional medicine. Indigenous societies have employed many plants in this family to heal various diseases for years. Local populations rely on these herbs for wound healing, pain treatment, and digestive issues [213]. Livestock ailments can be addressed with Nyctaginaceae species, which aid ethnoveterinary knowledge. This shows how deeply these plants affect rural populations. Nyctaginaceae plants benefit the environment ecologically [214].

Some species feed pollinators, while others support tiny animals. These plants also stabilize soil in arid locations and cycle nutrients, improving ecosystem health. In conclusion, the Nyctaginaceae family's extensive range, numerous uses in traditional medicine and ethnobotany, and ecological benefits make it meaningful in natural ecosystems and human communities.

• *Boerhaavia diffusa* L.

Local Name: Gajpurna

Botanical Characteristics: Diffused herbs; nodes hairy and thickened; leaves simple, opposite, often unequal; flowers in axillary or terminal panicles, purple; fruits oblong.

Distribution: Common weed found everywhere, has high adaptability from wet to extreme dry conditions.

Therapeutic Usage:

Ethnopharmacology:

• Leaf paste with ginger is given to children in the treatment of liver and spleen enlargement.
• Plant extract with goat milk is given to children to cure night blindness.
• Root decoction is used to treat jaundice. 1 teaspoon aqueous extract of roots with black pepper is given thrice a day for seven days. Roots are also tied around the neck of the patient.
• Cooked leaves are taken orally to check body inflammation.
• Leaf juice with black pepper is given as an antidote to snake bite.

• *Mirabilis jalapa* L.

Local Name: Nakta, Gulabaans

Botanical Characteristics: Herbs; leaves opposite, simple; flowers in racemes, tubular, various coloured; fruits nuts, ellipsoid, rugose.

Distribution: Common in waste places and near the villages.

Therapeutic Usage:

Ethnopharmacology:

• Root tuber is ground with jaggery and two spoonfuls is given once a day for a week to cure sprains.

- Tuberous root paste of white-flowered variety is applied locally on scorpion stings.
- The root powder is prescribed as a laxative.
- Tribals apply green leaves on boils.

FAMILY OLEACEAE

Introduction

The Oleaceae family, a botanical group of great ecological and cultural importance, includes many plant species worldwide. This diverse family is known for its ecological contributions, traditional medicine, ethnobotanical practices, and ethnoveterinary care [189]. Oleaceae plants grow in temperate and tropical habitats. Olive, jasmine, and lilac trees are famous members of this family [215]. They flourish in Mediterranean, subtropical, and temperate regions. The Oleaceae family is essential in traditional medicine and ethnobotany. Various species in this family have been used medicinally for ages. Olive oil, extracted from olive tree fruit, is famous for its culinary and medicinal purposes [216]. Jasmine has also been praised for its scent and healing properties. Due to ethnobotanical expertise, these plants are used in cosmetics, fragrance, and religious ceremonies [217]. Some Oleaceae species are used in ethnoveterinary medicine to treat livestock and pets. These traditional practices show rural people the practical and cultural value of these plants. Oleaceae supports biodiversity ecologically [218]. Some species support local ecosystems by providing nectar and pollen for bees and butterflies. Their versatility and capacity to grow in different settings make them valuable in natural and cultivated landscapes. The Oleaceae family's range, traditional usage in medicine and ethnobotany, and ethnoveterinary contributions demonstrate its diverse value. Its ecological importance in wildlife and ecosystems makes it essential in natural and human environments.

• *Nyctanthes arbor-tristis* L.

Local Name: Chirata

Botanical Characteristics: Small trees; branches quadrangular; leaves opposite, simple, ovate, rough; flowers in trichotomous cymes, white with bright orange tubes, fragrant; fruits capsules, flat, broad at apex; seeds flat.

Distribution: Common as self-sown and planted in forest areas and village sides.

Therapeutic Usage:

Ethnopharmacology:

- The fresh basal part of the bark is chewed, and the saliva is held inside the mouth for 3-5 minutes. This is repeated frequently at an interval of 4 hours for two days to cure teeth infections and troubles.
- In malarial fever, decoction of leaves and black pepper is given thrice a day for 3-4 days.
- Dried seed powder with mustard oil is applied to the scalp to remove dandruff.
- Seed paste is applied on piles.

FAMILY OXALIDACEAE

Introduction

Oxalidaceae, a varied and internationally dispersed family of flowering plants, has a rich botanical diversity, traditional applications, and ecological impacts. This family is essential in medicine, ethnobotany, and ethnoveterinary practices worldwide [219]. Oxalidaceae plants thrive in temperate and tropical climates. They are found in the Americas, Africa, Asia, and Australia. This extensive distribution shows their adaptability to many habitats [220]. Medically several Oxalidaceae species are medicinal. Creeping wood sorrel, *Oxalis corniculata*, has been used in traditional medicine to treat digestive and skin issues [221]. Wood sorrel (*Oxalis acetosella*) was used to prevent scurvy in several civilizations due to its vitamin C content. *Averrhoa carambola*, the star fruit, is also used medicinally and culinary [222]. Oxalidaceae plants have many ethnobotanical uses they are used as food in local cuisine [175]. In ethnoveterinary care, these plants help livestock. Their usage in treating common livestock ailments shows their rural utility. Oxalidaceae species benefit the environment [189]. *Oxalis corniculata, Oxalis acetosella,* and *Averrhoa carambola* demonstrate the family's richness and many uses in human and natural surroundings [223].

• *Oxalis corniculata* L.

Local Name: Chuka buti

Botanical Characteristics: Herbs; stems trailing, rooting at nodes; leaves alternate, trifoliate; flowers in sub umbellate cymes, yellow; fruits capsules, cylindrical; seeds ovoid, brown, black, striate.

Distribution: Common as weed in moist places, in shades, and in cultivations.

Therapeutic Usage:

Ethnopharmacology:

• Fresh juice of plants cures dyspepsia and piles.
• An infusion of the leaves is used to remove the opacity of the cornea.
• Plant paste along with 'mahi' (butter-free curd) is taken to control blood in the stool.
• Leaf paste is applied to the head to cure insomnia.
• Leaf juice is given as an antidote to snake poison.

FAMILY PAPAVERACEAE

Introduction

The Papaveraceae family of flowering plants is known for its botanical diversity, traditional applications, and ecological importance. This family is essential in worldwide medicine, ethnobotany, and ethnoveterinary practices [224]. From temperate woods to arid deserts, Papaveraceae plants find extraordinary adaptability. They thrive in various settings in Europe, Asia, and North America. Pharmacologically, Papaveraceae is important. Many species in this family include alkaloids and other bioactive chemicals used in traditional medicine throughout history [225]. Opium (*Papaver somniferum*) contains painkillers like morphine and codeine. In Native American medicine, the bloodroot (*Sanguinaria canadensis*) may have antibacterial qualities [226]. The socio-botanical functions of these plants are varied. Their medicinal uses are supplemented by cultural rites and ceremonies symbolizing life and death. Papaveraceae species are less commonly used in ethnoveterinary medicine to manage animal health. The Papaveraceae family encourages biodiversity. Providing nectar and pollen to pollinators helps many plant species reproduce. For ecological benefits, the family is not as well known as other plant families [227].

• *Argemone mexicana* L.

Local Name: Pili Kataeyya

Botanical Characteristics: Prickly herbs, annual, glaucescent, laticiferous; leaves spinulose, dentate on the margins; flowers bright yellow; stigma lobed, oppressed.

Distribution: Very common as a weed in waste places and in cultivations.

Therapeutic Usage:

Ethnopharmacology:

- Leaf juice is mixed with rhizome paste of *Acorus calamus*. This is applied locally on the eczema-affected part.
- The whole part along with the rhizome of *Curcuma longa* is taken in equal proportions and burnt into ash. This is mixed with coconut oil made into a paste and applied continuously on psoriasis-affected part for about a month.
- Tribals use plant juice in the treatment of eye troubles like conjunctivitis.

Ethnoveterinary:

Tribals apply plant juice with a paste of onion to kill parasitic insects on the bodies of pet animals.

FAMILY PIPERACEAE

Introduction

Piperaceae, a large and internationally dispersed family of flowering plants, is rich in botanical diversity, traditional usage, and ecological value [232]. This global family is essential in medicine, ethnobotany, and ethnoveterinary practices. Tropical rainforests and subtropical climates suit Piperaceae plants. They are found throughout South America, Southeast Asia, and Africa, demonstrating adaptability [233]. For healthcare, the Piperaceae family is crucial. Many species in this family include bioactive substances, including alkaloids and essential oils used in traditional medicine for ages [234]. *Piper nigrum,* the black pepper plant, is famous for its culinary and medicinal purposes. In Pacific Island cultures, *Piper methysticum* (kava) is used for its anxiolytic and sedative properties [235, 236]. Piperaceae plants have many ethnobotanical uses. Ethnoveterinary treatment can employ Piperaceae species to manage animal health [237]. However, such practices are region-specific and less documented than other plant families. Piperaceae species support biodiversity by providing habitat and food for insects and birds. Their roles include nutrient cycling and ecosystem function. Piperaceae species include *Piper nigrum* (black pepper), *Piper methysticum* (kava), and *Piper aduncum*. These species demonstrate the family's botanical richness and historical and current usefulness in human cultures and ecosystems [238].

- *Piper longum* L

Local Names: Pipal, Piparmul

Botanical Characteristics: Aromatic herbs; branches swollen at nodes; leaves alternate, simple, ovate; flowers in spike, cylindrical, unisexual; fruits ovoid, globose.

Distribution: Common as a wild stage in open areas within dense forests.

Therapeutic Usage:

Ethnogynaecology:

Tribals use roots, about 3 cm long, to cause abortion up to 3-4 months.

Ethnopharmacology:

• Powder of dried fruits along with butter-free curd is given for the treatment of gas ball (Vayu-gola).
• Green fruits are advised to cure nasal congestion, whooping cough, *etc.*
• A mixture of pipal, cardamom, madhu, and *Centella* is prepared in jaggery and given for chronic asthma.
• Ripe sun-dried fruits are powdered and taken twice a day for intestinal worms.

Ethnoveterinary:

• Besides human usage, it is also administered in common cold problems of their pet animals.

FAMILY POACEAE

Introduction

One of Earth's most ecologically and economically important plant families is the Poaceae, or grass family. With thousands of species in many environments, Poaceae is critical to agriculture, ecology, medicine, and traditional practices [239]. Nearly every continent is home to Poaceae. These grasses can grow in freezing Arctic tundra, searing Sahara deserts, and lush Amazon rainforests [240]. Their widespread presence in terrestrial habitats emphasizes their ecological relevance. Poaceae is best known for producing rice, wheat, and maize, although numerous species have therapeutic properties [241]. Bamboo (*Bambusa spp.)* can alleviate fever and inflammation, according to traditional Asian medicine.

The diuretic couch grass (*Elymus repens*) has also been used in herbal medicine [242]. Poaceae plants have many ethnobotanical uses beyond food. Indigenous tribes have used grass throughout to weave baskets, make tools, and build shelters. In ethnoveterinary practices, specific grasses can help animals [243]. For livestock digestive difficulties, Bermuda grass (*Cynodon dactylon*) is employed.

Poaceae species underpin terrestrial ecosystems. They feed and shelter many herbivores and small mammals. They cycle nutrients and avoid soil erosion through their enormous root systems [244]. Poaceae-dominated grasslands are important carbon sinks, mitigating climate change. Bamboo, soft grass, and Bermuda grass are Poaceae species. These plants demonstrate the family's botanical richness and wide range of impacts on societies, ecosystems, and food security.

• *Bambusa stricta* Roxb.

Local Name: Bans-dhan

Botanical Characteristics: Densely caespitose bamboos; branches with thorns; leaves alternate, simple, linear-lanceolate; flowers in panicle, spikelet with many flowers; grains linear-oblong.

Distribution: Common wildly in forests as well as cultivated.

Therapeutic Usage:

Ethnogynaecology:

Crystals obtained from internodes called 'Banslochan' with a decoction of 7 long peppers are given to women in the treatment of constitutional disorders.

Ethnopharmacology:

• 10 gm of 'Banslochan' is given to cure acidity.
• Tabashir, a liquid substance accumulated in internodes, is prescribed for the treatment of asthma.
• Green epidermal cells are applied on new cuts to stop bleeding.
• A paste of young leaves with a paste of ginger is prescribed for the treatment of diabetes.

Ethnoveterinary:

• Tribals give leaves to the cattle as fodder for quick expulsion of the fetus.

FAMILY RHAMNACEAE

Introduction

The Rhamnaceae family of flowering plants is remarkable because it combines botanical variety, traditional applications, and ecological significance. Rhamnaceae is a family of plants with essential functions in medicine, ethnobotany, and even ethnoveterinary practices in many parts of the world [245]. Plants of the family Rhamnaceae can thrive in a wide range of habitats, from deserts to moist forests. Their widespread existence globally, from North and South America to Europe, Asia, and Africa, attests to their adaptability and resilience in various environments [246]. The Rhamnaceae family has been used for centuries as a source of medicine. Buckthorn *(Rhamnus cathartica)* is just one member of this family that has a long history of being used as a laxative and for its potential to promote digestive health. Some, like jujube (*Ziziphus jujuba*), have been used medicinally for centuries [247]. Jujube is often used to treat insomnia and anxiety. The Rhamnaceae family of plants has many applications in ethnobotany. Their cultural value is displayed in the communities where they are employed by making tools, baskets, and dyes [248]. While regional and minor research, these plant families may have potential uses in ethnoveterinary medicine for treating and managing animal illness. Rhamnaceae plants are essential to the environment because they provide shelter and food for numerous birds and insects [249]. In dry environments, they help to reduce soil erosion and promote nutrient cycling through their activities in stabilizing soil. Buckthorn (*Rhamnus cathartica*), Jujube (*Ziziphus jujuba*), and Chaparral (*Larrea tridentata*) are all Rhamnaceae species. These examples of plant diversity highlight the family's historical and modern significance in human cultures, ecosystems, and rituals.

• *Zizyphus nummularia* (Burm. F.) W & A.

Local Name: Jhadber

Botanical Characteristics: Large erect shrubs; up to 2 m high with sharp thorns on long sarmentose branches; leaves 3-5 cm long, obliquely ovate-lanceolate; flowers in dichotomous cymes, greenish-yellow; fruits drupe.

Distribution: Common in shady forests and wastelands.

Therapeutic Usage:

Ethnopharmacology:

• Fresh leaf paste is used as a laxative and given for throat troubles.
• Stem bark decoction with long pepper paste is given to cure cough, cold, and fever.
• Leaf paste with long pepper is prescribed as an antidote to chicken pox.
• Stem bark decoction is extensively used in the washing of sore throats.

FAMILY RUBIACEAE

Introduction

Rubiaceae's flowering plant family is significant and widespread, and its members offer a rich tapestry of botanical variety, traditional applications, and ecological value. The family Rubiaceae, which has more than 13,000 species distributed around the globe, is vitally essential in fields as diverse as medicine, ethnobotany, and even ethnoveterinary practices [250]. Plants in the family Rubiaceae can thrive in various habitats, from tropical rainforests to temperate deciduous woods. Their widespread prevalence across Africa, Asia, the Americas, and even some sections of Australia attest to their adaptability to various environmental conditions [251]. The Rubiaceae family has been used for centuries as a source of medicine. Several members of this family have been used medicinally for centuries due to the presence of bioactive chemicals like alkaloids and antioxidants. The quinine tree, or *Cinchona officinalis*, is well-known for its traditional usage in malaria treatment [252]. The coffee plant (*Coffea arabica*) is also lauded for its ability to stimulate the brain and nervous system, thanks to its high caffeine content [253]. The plants of the family Rubiaceae are put to many different applications in ethnobotanical contexts. They are frequently used in rituals and celebrations, symbolizing diverse aspects of local communities' lives and spiritualities. Some species of Rubiaceae may have uses in ethnoveterinary care for animal health management. However, these uses tend to be regional and little documented compared to those of other plant families [254]. Species in the family Rubiaceae support biodiversity by providing shelter and sustenance for numerous animal species, including numerous bird and insect species. Mainly in tropical rainforests, the presence of certain species is crucial for maintaining the integrity of the soil. Rubiaceae family includes the quinine tree (*Cinchona officinalis*), the coffee plant (*Coffea arabica*), and the cape jasmine (*Gardenia jasminoides*). In this way, the richness of this plant family is highlighted, as is the family's history and contemporary significance to human communities, ecosystems, and traditional practices [255].

• *Xeromphis spinosa* (Thumb.) Keay

Local Name: Maini

Botanical Characteristics: Large shrubs or small trees with axillary spines; leaves 3-5 cm long, ovate, obovate, obtuse or acute; flowers yellowish white, fragrant; fruits soft, fleshy berry, globose or ovoid, greenish yellow.

Distribution: Common along the cultivated fields, waste places, and forest areas.

Therapeutic Usage:

Ethnopharmacology:

- Fruit pulp is given to treat dysentery and diseases of the urinary tract.
- Extract of fruits is administered as insecticidal.
- Fruit decoction is prescribed as a blood purifier.
- Root decoction with long pepper is given to treat nervous disorders.

FAMILY RUTACEAE

Introduction

There is a rich tapestry of botanical variety, traditional applications, and ecological value to be found among the members of the Rutaceae family of flowering plants. The Rutaceae family, which includes over 160 genera and more than 2,000 species, is essential in medicine, ethnobotany, and even ethnoveterinary practices [256]. Plants in the family Rutaceae are exceptionally versatile, able to survive, and even thrive in a wide range of environments, from humid tropical forests to dry, sandy deserts. Their widespread occurrence in regions as different as Asia, the Americas, Australia, and Africa attests to their adaptability to various environments [257]. The Rutaceae family of plants is highly significant in the field of medicine. Bioactive substances like alkaloids and essential oils are found in many species in this family and have been used for ages in alternative medicine. The lemon tree (*Citrus limon*) is lauded for its health benefits due to its fruit's abundance of vitamin C [258]. Similarly, rue (*Ruta graveolens*) has been used traditionally as an emmenagogue and antispasmodic, giving it potential therapeutic value. Plants of the family Rutaceae have several applications in ethnobotany [259]. They have many symbolic meanings in local cultures and are frequently used in rituals and ceremonies related to life, safety, and cleansing. Although regional and less well-documented than other plant families, some Rutaceae plants may find benefits in ethnoveterinary care for regulating animal health [260]. Species of the family Rutaceae benefit ecosystems because they provide shelter and food for various animals, including insects and birds. Particularly in their native environments, certain species also aid in soil stabilization and nutrient cycling.

- *Aegle marmelos* (Linn.) Correa

Local Name: Bilva

Botanical Characteristics: A small deciduous glabrous tree, thorny; leaves pale green, trifoliate, lateral leaflets sessile, ovate-lanceolate, 3-5 inches long, terminal, long-petioled; flowers greenish white, an inch in diameter, sweetly scented; fruits globose, hard.

Distribution: Scattered, near the villages, cultivated in the fields or the forests.

Therapeutic Usage:

Ethnopharmacology:

- Ripe fruit pulp is given to drink in the morning or powder of half-ripe fruit is given twice daily for 2-3 days to cure digestive disorders, piles, *etc.* and to control diarrhea and dysentery.
- Raw pulp alone or its drink is prepared in water and is taken to treat heat stroke.
- Half-ripe fruit is used as an astringent, digestive, and stomachic.
- Its leaves with neem leaves and Tulsi leaves are dried and powdered which is given for treatment and control of diabetes.

FAMILY SOLANACEAE

Introduction

The Solanaceae family, which includes many valuable plants, is widespread around the globe and exhibits an intriguing synthesis of botanical variety, cultural relevance, and ecological importance. The Solanaceae family, which has over 100 genera and more than 2,500 species, is essential in many human and animal health areas [261]. Plants in the Solanaceae family are incredibly versatile and can survive in various settings, from humid tropical forests to dry, arid regions. Their widespread occurrence around the globe, especially in the Americas, Asia, and Africa, attests to their adaptability to a wide range of environments. The Solanaceae family of plants is highly significant in the field of medicine. Alkaloids and glycoalkaloids, found in many species in this family, have been used for millennia in alternative medicine [262]. *Atropa belladonna*, or deadly nightshade, has been used as a sedative and pain reliever in the past, and this fact lends historical weight to the plant's possible medical benefits [263]. The chili pepper plant, *Capsicum annum*, is also revered for its medical and culinary applications, with many people turning to it for the capsaicin in the peppers [264]. The Solanaceae family of plants has many applications in ethnobotany. They are commonly used in ceremonies and rituals and represent important meanings in local religions and cultures. Although regional and less well-documented than other plant families, some plants of the Solanaceae family may find applications in ethnoveterinary care for regulating animal health [264]. Solanaceae plants are essential to biodiversity because they provide shelter and food for many animals. Certain species' allelopathic effects may affect nearby plants' growth and competitiveness. The deadly nightshade (*Atropa belladonna*), the chili pepper (*Capsicum annuum*), and tobacco (*Nicotiana tabacum*) are all members of the Solanaceae family. These examples of plant diversity highlight the family's historical and modern significance in human cultures, ecosystems, and rituals.

• *Solanum torvum* Swartz

Local Name: Khat-khiri

Botanical Characteristics: Shrubs; leaves alternate, simple, prickly, ovate, elliptic, lobed; flowers in racemes, white; fruits berries, globose; seeds orbicular, compressed, smooth, and black.

Distribution: Common in bushes and in wastelands.

Therapeutic Usage:

Ethnopharmacology:

• Fumes of burning seeds are inhaled for toothache.
• Dried fruit powder is given to an enlarged liver.
• Fruits are given to children with coughs and cold.

• *Solanum nigrum* L.

Local Name: Makoiya

Botanical Characteristics: Erect, glabrous, unarmed annual herbs; leaves ovate-oblong, 6-9x2 cm, toothed or lobed, sparsely pubescent; flowers white, in drooping umbellate extra axillary cymes; berries globose, glabrous.

Distribution: Common in bushes and in wastelands.

Therapeutic Usage:

Ethnogynaecology:

• About 10 gm of root powder is given with water to women on 4th, 8th and 12th day after the Menstrual Cycle for successful pregnancy.

Ethnopharmacology:

• Leaf juice is applied in bee and wasp bites.
• Ripen fruits are eaten by children for immunity boosters.

• *Withania somnifera* (L.) Dunal

Local Name: Asgandh

Botanical Characteristics: Shrubs; leaves alternate, simple, ovate; flowers in axillary cymes, greenish yellow; narrow-mouthed; fruits berries, red; seeds small.

Distribution: Not very common but occurs wild.

Therapeutic Usage:

Ethnogynaecology:

• Seed decoction or seed paste is given to women for amenorrhea and other female disorders.

Ethnopharmacology:

• Seed paste is applied to skin diseases.
• Seed powder is recommended as a diuretic and narcotic.
• Dried root powder with goat milk is given to cure general sex debility and vigor.

FAMILY VERBENACEAE

Introduction

Verbenaceae, a complex and internationally dispersed family of flowering plants, combines botanical diversity, traditional usage, and ecological relevance. Verbenaceae, with 100 genera and over 3,000 species, is important in medicine, ethnobotany, and ethnoveterinary practices [265]. Verbenaceae thrive in tropical rainforests and arid deserts. Their widespread distribution across the Americas, Africa, Asia, and even parts of Australia show their adaptability to varied habitats. The Verbenaceae family is important medicinally. Many species in this family include bioactive substances, including essential oils and alkaloids used in traditional medicine for ages [266]. *Verbena officinalis*, or vervain, is known for its mild sedative and anti-inflammatory qualities [267]. While invasive in some areas, *Lantana camara* has been used in herbal medicine for its antibacterial and anti-inflammatory effects [268]. Verbenaceae plants support biodiversity by providing habitat and food for insects, birds, and butterflies. Some organisms help stabilize soil and cycle nutrients in tropical habitats. Verbenaceae species include *vervain, lantana*, and wild jasmine. These species demonstrate the family's botanical richness and historical and contemporary importance in human communities, ecosystems, and traditional practices.

• *Vitex negundo* L.

Local Name: Ratalu

Botanical Characteristics: Shrubs or small trees; branches quadrangular, densely white-tomentose; leaves 3-5 foliate; leaflets stalked, lanceolate, acuminate, entire; flowers blue, in loose clusters, arranged in large terminal panicles; drupes black.

Distribution: Occasionally planted near village sides.

Therapeutic Usage:

Ethnopharmacology:

• Green leaves are chewed in cough, associated cold, and fever. Leaf juice is given three times a day for three days.
• Fresh flower extract is administered in diarrhea.
• Plant extract is used to expel worms in children.
• Seed oil is massaged over the joints to get relief from arthritis.

FAMILY ZINGIBERACEAE

Introduction

Zingiberaceae, a rare and widespread family of flowering plants, is fascinating for its botanical diversity, traditional applications, and ecological importance. Zingiberaceae, with 53 genera and 1,300 species, is important in medicine, ethnobotany, and ethnoveterinary practices [269]. Zingiberaceae plants thrive in tropical rainforests, subtropical regions, and temperate climes. They are abundant throughout Asia, the Americas, and the Pacific Islands, demonstrating their adaptability. The Zingiberaceae family is vital medicinally. Many species in this family produce bioactive chemicals, including gingerols and curcuminoids, which have been utilized in traditional medicine for millennia. Ginger, or *Zingiber officinale*, reduces nausea and inflammation [270]. *Curcuma longa*, the turmeric plant, is recognized in traditional medicine for its curcumin and anti-inflammatory and antioxidant properties [271]. Zingiberaceae species support biodiversity by providing habitat and food for insects and birds. In tropical settings, certain species enhance soil and cycle nutrients. *Alpinia galanga, Zingiber officinale,* and *Curcuma longa* are possibilities for Zingiberaceae species. These species demonstrate the family's botanical richness and historical and contemporary importance in human communities, ecosystems, and traditional practices.

• *Curcuma longa* L.

Local Name: Haldi

Botanical Characteristics: Perennial herb; a short, thickened rhizome bearing a tuft of large, broad, lanceolate leaves; pale yellow flowers are borne in dense spikes terminating the stem; flowers remain covered by pink bracts.

Distribution: Cultivated in fields and kitchen gardens.

Therapeutic Uses:

Ethnopharmacology:

• The rhizome is given in diarrhea and fevers.
• Fresh juice of rhizome is given as anthelmintic.
• Fresh rhizome juice is also used to clean foul ulcers and skin affections.
• The rhizome powder is applied on cuts and wounds as an antiseptic.

• *Zingiber officinale* Rosc.

Local Name: Adrak

Botanical Characteristics: Perennial herbs; leaves lanceolate-oblong, punctuate above, appressed hairy beneath; corolla tube pubescent.

Distribution: Cultivated in fields and kitchen gardens.

Therapeutic Usage:

Ethnopharmacology:

• Juice mixed with honey is taken for cough and cold and also for the expulsion of intestinal worms.
• Fresh rhizome pieces with salt are kept in the mouth to relieve cough.
• Dried powder is taken with water for arthritis pain.

• *Zingiber roseum* (Roxb.) Roscoe.

Local Name: Banabyada

Botanical Characteristics: Perennial herbs; leaves lanceolate-oblong, punctuate above, oppressed hairy beneath; corolla tube pubescent.

Distribution: Frequent on moist slopes or rocks, cultivated.

Therapeutic Usage:

Ethnopharmacology:

- Fresh leaf paste is prescribed to treat cough and cold.
- Tribals apply rhizome paste with long pepper paste on the fractured bone.
- Rhizome is used as an aromatic, carminative, and digestive agent.

CONCLUSION

In summary, the research conducted on the above-mentioned plant families in Pilibhit, recognized as a biodiversity hotspot in the Indo-Nepal Terai Arc Landscape, thoroughly investigates their botanical characteristics, historical applications, and ecological significance. Globally Pilibhit Tiger Reserve PTR is situated in a distinctive geographical position in the Northern Hemisphere within the sub-tropical climate zone which encompasses a wide range of ecosystems, making it an optimal place for doing research on plant biodiversity. The study explores the utilization of medicinal plants and commercially important species, providing insights into their geographical distribution, ecological functions, and cultural importance. The results demonstrate a complex network of connections between these botanical families and the surrounding community. This study reveals the medicinal, ethnobotanical, and ethnoveterinary applications of these plants, highlighting their significant influence on local healthcare and traditional customs (Table **1**).

Moreover, the study delves into the ecological services provided by these organisms, including their contribution to habitat maintenance, provision of food resources, and promotion of soil stability. This highlights their pivotal function in upholding regional biodiversity and the overall functioning of ecosystems. This study highlights the need to conserve the botanical diversity of Pilibhit Tiger Reserve and the valuable resources contained within these plant groups. In addition to its role in documenting information, the study places significant emphasis on the practical implications for conservation efforts in Tran boundary Landscape-crucial for species movement, sustainable resource management, and the preservation of traditional knowledge. It is decisive to acknowledge the complex interdependencies between humans and plants within this region of high biodiversity to promote the preservation of biodiversity and improve the overall welfare of the community. This study serves as a pivotal resource for making well-informed decisions that seek to achieve a balance between human activities and the intricate ecosystems of Pilibhit Tiger Reserve, which forms a vital corridor for wildlife movement across international boundaries, with the goal of benefiting both the natural environment and its residents.

Table 1. Enumeration of Medicinal Plants of Pilibhit Tiger Reserve, PTR.

S. No.	Botanical Name of the Plant	Local Name	Family	Medicinal Uses	Specific Attributes
1	Justicia gendarussa	Adusa	Acanthaceae	In respiratory disorders, fever and digestive troubles.	Leaves
2	Ruellia tuberosa	Bhukanda	Acanthaceae	In diuretic disorders.	Tubers
3	Barleria prionitis	Vajradanti	Acanthaceae	Used externally on wounds and skin problems.	Leaves
4	Adhatoda vasica	Vasaka	Acanthaceae	Cough, cold, and tuberculosis.	Leaves
5	Peristrophe bicalyculata	Hadjor	Acanthaceae	Fractured bones, eczema treatment, *etc.*	Leaves
6	Amaranthus virdis	Chaulai	Amaranthaceae	General weakness, malnutrished patients.	Leaves
7	Alternanthera sessilis	Grundi	Amaranthaceae	Anti-inflammatory and anti-oxidant.	Leaves
8	Gomphrena globose	Gol manika	Amaranthaceae	Fluid retention, edema patients.	Leaves
9	Achyranthes aspera var aspera	Chinchilla	Amaranthaceae	Contraceptive, snake bite,mad-dog bite.	Twig
10	Achyranthes aspera var porphyristachia	Chirchiri	Amaranthaceae	Mental disorders, asthma.	Leaves
11	Magnifera indica	Aam	Anacardiaceae	digestive disorders, and diuretic properties.	Fruits
12	Buchaniana lanzan	Kath Bhilwa	Anacardiaceae	Rheumatic pain, glandular pain.	Stem bark
13	Semecarpus anacardium	Bhailwar	Anacardiaceae	Contraceptive, rheumatism.	Root/ Seeds
14	Daucus carota	Gajar	Apiaceae	Digestive troubles, vitamin B source.	Root
15	Apium graveolens	Ajwain patta	Apiaceae	Blood pressure maintenance, Vitamin A Source.	Leaves
16	Coriandrum sativum	Dhaniya	Apiaceae	Digestive troubles, Kidney purification.	Leaves
17	Petroselinum crispum	Ajmoda	Apiaceae	Diuretic characteristics	Leaves
18	Foeniculum vulgare	Saunf	Apiaceae	Digestive disorders and menstrual ailments.	Seed
19	Centella asiatica	Brahmi booti	Apiaceae	mental disorder and as a brain tonic.	Flowers
20	Hemidesmus indicus	Gorkatala	Apocynaceae	Leucoderma, dyspepsia, blood purification *etc.*	Roots
21	Nerium oleander	Lal kaner	Apocynaceae	In skin problems	Latex
22	Vinca minor	Sadauli	Apocynaceae	To enhance cognitive function and memory.	Leaves
23	Thevetia peruviana	Peeli kaner	Apocynaceae	Skin problems	Latex
24	Alstonia scholaris	Chatwan	Apocynaceae	Expulsion of helminths from the intestine, application of latex to cure caries of teeth.	Stem bark/Latex *etc.*

(Table 1) cont.....

S. No.	Botanical Name of the Plant	Local Name	Family	Medicinal Uses	Specific Attributes
25	Carissa opaca	Jangli karonda	Apocynaceae	Cough, cold and rheumatoid fever, *etc.*	Fruits/ Stem decoction
26	Catharanthus roseus	Sadabahar	Apocynaceae	Diabetes, insect bites *etc.*	Leaves
27	Holarrhena antidysenterica	Duddhi	Apocynaceae	To minimize labor pain in women.	Stem bark decoction
28	Rauvolfia serpentina	Naag bel	Apocynaceae	Snake bite, hypertension and high B.P *etc.*	Root
29	Colacasia esculenta	Ghuiyaan	Araceae	Detoxification and in gastronmic preparation.	Corn
30	Syngonium podophyllum	Gusoot	Araceae	Air purification and wound healing and bacterial infection *etc.*	Leaves
31	Epipremnum aureum	Money Plant	Araceae	Air purification anti-bacterial, anti-fungal, anti-inflammatory, anti-cancer, and anti-termite activity.	Leaves/ Root
32	Acorus calamus	Papari	Araceae	Chickenpox, bronchitis, cold cough,Fever *etc.*	Root/Rhizome
33	Aclepias syriaca	Duddhi	Asclepiadaceae	pain relief, kidney and urinary dysfunction, rheumatism and backache, antioxidants *etc.*	Root/ Latex
34	Ceropegia woodii	Dil bel	Asclepiadaceae	Air purification, anti-bacterial,anti-allergic, combat anxiety, *etc.*	Leaves
35	Hoya carnosa	Mome paudha	Asclepiadaceae	Anti-microbial, air purification, diabetes, kidney and urinary problems.	Leaves
36	Asclepias curassavica	Kak tund	Asclepiadaceae	Stops bleeding, skin-warts, pneumonia, diarrhea, abortifacient, *etc.*	Root/plant decoction
37	Calotropis gigantea	Akauwa	Asclepiadaceae	Abdominal pain, toothache,thorns pull out,rhematism *etc.*	Leaves
38	Helianthus annus	Suryamukhi	Asteraceae	Fewer sore throat, snake bites, pulmonary affections, cough, and cold, malaria, heart troubles, antioxidants, analgesic, *etc.*	Leaves/ Seeds
39	Taraxacum officinale	Singhparni	Asteraceae	Hepato protection and digestive properties	Root/ Leaves
40	Calendula officinalis	English genda	Asteraceae	Anti-inflammatory, skin rashes and burns,skin care *etc.*	Flowers
41	Chrysanthemum indicum	Guldaudi	Asteraceae	Anti-inflammatory, anti-microbial,anti-oxidant, respiratory disorders, hypertension *etc.*	Flowers
42	Elephantopus scaber	Bishari	Asteraceae	Eczema, fever, vomiting, filarial swellings,insect bites	Leaves/ Root
43	Veronia anthelmintica	Kalijiri	Asteraceae	Anthelmintic, piles, dysentery, leucuderma, etc	Seeds
44	Oroxylum indicum	Sauna	Bignoniaceae	Stomach-ache, rheumatic swellings, diarrhea and dysentery, bone fracture, *etc.*	Root bark/ Stem bark
45	Adansonia digitata	Gorakh imli	Bombacaceae	Anti-inflammatory,analgesic,anti-oxidant,anti-microbial,anti-pyretic properties	Seeds/ Fruit pulp
46	Litsea speciosa	Resham rui	Bombacaceae	anti-bacterial,anti-inflammatory,anti-pyretic,Hepato protective,anti-oxidants *etc.*	Seed oil/Stem bark
47	Ceiba pentandra	Safed semal	Bombacaceae	Anti-acne,anti-microbial,skin problems like chicken pox, anti-aging	Leaves/ Bark
48	Pachira aquatica	Money Plant	Bombacaceae	Stomach problems, anemia, high BP,f atigue and debility	Stem bark/ Seeds/ Leaves
49	Bombax ceiba	Simra	Bombacaceae	Diarrhea, small pox, dysmenorrhea *etc.*	Flowers/ Stem bark
50	Cordia dichotoma	Lasuara	Boraginaceae	Dysmenorrhea,purgative, cough and cold, diabetes	Stem bark/Fruits/ Root

(Table 1) cont.....

S. No.	Botanical Name of the Plant	Local Name	Family	Medicinal Uses	Specific Attributes
51	Caesalpinia crysta	Khaja	Caesalpiniaceae	Stomach ache,facial paralysis, malarial fever, dyspepsia *etc.*	Leaves/ Seeds
52	Cassia fistula	Sinara	Caesalpiniaceae	Jaundice, septic wounds, leprosy, whooping cough *etc.*	Root/ Stem
53	Cassia tora	Chakwar	Caesalpiniaceae	Bone fracture, eczema, wound healing, malarial fever	Bark/ Fruits
54	Cassia occidentalis	Kasondha	Caesalpiniaceae	Diabetes, ringworms ringworms, scabies, *etc.*	Root/bark/ Seeds
55	Capparis zeylanica	Zakhambela	Capparidaceae	Tuberculosis, hydrocoele, cholera, smallpox *etc.*	Root/ Stem/ Bark
56	Terminalia alata	Asana	Combretaceae	Pneumonia, leprosy, *etc.*	Stem bark
57	Terminalia arjuna	Kahua	Combretaceae	Chronic fever, teeth problems,ear-ache *etc.*	Leaves/ Stem bark
58	Terminalia bellerica	Baheda	Combretaceae	Vomiting and lose motion, fever, cough, *etc.*	Fruits
59	Terminalia chebula	Harra	Combretaceae	Diarrhea, flatulence, eyes inflammation	Fruits
60	Momordica charantia	Karela	Cucurbitaceae	Heat stroke, diabetes	Leaves/fruits
61	Cuscuta reflexa	Sarag Baburi	Cuscutaceae	Joint dislocation, swollen testicles, jaundice, headache, *etc.*	Twigs/ Seeds
62	Dioscorea bulbifera	Belarkanda	Dioscoreaceae	abortion, bone fractures, sores *etc.*	Tubers
63	Emblica officinale	Amla	Euphorbiaceae	Indigestion,constipation,scurvy diarrhea, jaundice *etc.*	Fruits
64	Jatropha curcas	Bakrenda	Euphorbiaceae	Eczema, teeth cleaning,tumors,bleeding *etc.*	Latex/ Seeds
65	Mallotus phillipensis	Rohini	Euphorbiaceae	Rheumatism,gall bladder stones, skin deases, boils, and blisters	Root/ Fruits
66	Phyllanthus fraternus	Jar amla	Euphorbiaceae	Genito-urinary problems, jaundice,dysentery *etc.*	Roots
67	Putranjiva roxburghii	Pinpina	Euphorbiaceae	Abortion, fever, *etc.*	Leaves/ Fruits
68	Abrus precatorious	Gughuchi	Fabaceae	White discharge, contraceptive, abortion, goiter, *etc.*	Root/ Seeds
69	Butea monosperma	Dhaka	Fabaceae	Bone fracture, testicular inflammation,night blindness,sunstroke.	Stem bark/ Seeds
70	Salvia officinalis	Samudraphal	Lamiaceae	Anti-oxidant, anti-inflammatory,anti-bacterial properties, dental problems, menopause symptoms, *etc.*	Leaves
71	Thymus vulgaris	Ban ajwain	Lamiaceae	Anti-bacterial,anti-oxidant, and anti-inflammatory characteristics, respiratory ailments, *etc.*	Whole plant
72	Ocimum basilicum	Dauna	Lamiaceae	Sexual weakness, fever, cough, cold, insect bites, *etc.*	Leaves/ Stem
73	Litsea glutinosa	Maida Lakri	Lauraceae	rheumatism, gout, bone fracture, *etc.*	Stem
74	Leea indica	Chattri	Leeaceae	Anti-inflammatory, joint discomfort and traumas	Leaves
75	Gluta usitata	Amili	Leeaceae	Gastro-intestinal ailments, astringent	Stem
76	Leea guineas	Bahasa	Leeaceae	Anti-inflammatory effects, arthritis, muscular discomfort, pain management	Leaves
77	Leea sambucina	Berry	Leeaceae	Diuretic properties, edema, *etc.*	Leaves
78	Leea asiatica	Golkandra	Leeaceae	Pneumonia, urinary problems, ringworm, etc	Leaves
79	Leea mycrophyll a	Lathigaja	Leeaceae	Better lactation, diarrhea, bone fracture, *etc.*	Roots

(Table 1) cont.....

S. No.	Botanical Name of the Plant	Local Name	Family	Medicinal Uses	Specific Attributes
80	Gustavia augusta	Kumbhi	Lecythidaceae	Anti-inflammatory, analgesic, etc	Leaves/ Stem bark
81	Barringtonia acutangula	Sunaffar	Lecythidaceae	Diarrhea, headache, toothache, *etc.*	Leaves/ Seeds
82	Aloe barbadensis	Gwar ka phatta	Liliaceae	Skin troubles, burns, wounds, digestive problems, *etc.*	Leaves
83	Polygonatum multiflorum	Satabari	Liliaceae	Anti-inflammatory properties, joint, muscle pain, *etc.*	Rhizome
84	Urginea maritima	Samudri pyaaz	Liliaceae	Cardiac and expectorant properties	Leaves
85	Allium cepa	Pyaaz	Liliaceae	Jaundice, spleen, digestive problems, flatulence and dysentery	Bulb
86	Allium sativum	Lassun	Liliaceae	Arthritis, digestive troubles,anti-aging, rheumatism, *etc.*	Cloves
87	Asparagus racemosus	Satavar	Liliaceae	Lactation in mothers, ulcer protection, Epilepsy, *etc.*	Root
88	Urginia indica	Patalu	Liliaceae	Leurcorrhea, jaundice, antidote for snake venom, bronchial troubles	Bulb/ Corm
89	Strychnos nux-vomica	Kuchila	Loganiaceae	Snake bite, antidote for poison, *etc.*	Root bark/ Stem bark
90	Lygodium flexuosum	Bartant	Lygodiaceae	Anaemia, diarrhea, stomach pain, *etc.*	Root
91	Lawsonia inermis	Mehandi	Lythraceae	Soar throat, skin disease, leprosy, etc	Leaf/ Flower
92	Abutilon indicum	Kanghi	Malvaceae	Scabies, abscess care, laxative properties	Leaves/ Seeds
93	Sida rhombifolia	Lal bariara	Malvaceae	Rheumatism, snake venom, etc	Root/ Leaves
94	Azadirachta indica	Neeba	Meliaceae	Chicken pox, fever, diabetes, etc	Root/ Leaves
95	Cissampelos pareira	Patat ki Bel	Menispermaceae	Abortion in pregnant women, snakebite, peptic ulcer, *etc.*	Leaves
96	Tinospora cordifolia	Gurch	Menispermaceae	Flu, fever, cold, cough, maintaining good health.	Leaves
97	Acacia catechu	Kharida	Mimosaceae	Wounds boils and burns, leprosy, *etc.*	Leaves
98	Albizzia lebbeck	Sarim	Mimosaceae	Septic,leprosy,antidote *etc.*	Stem/Seeds
99	Pithecellobium dulce	Jungle jalebi	Mimosaceae	Leucorrhea, intestinal worms, septic wounds	Root/ Leaves
100	Ficus racemosa	Gulariya	Moraceae	Lececorrhea, kidney stone, dysentery, *etc.*	Root/ Leaves
101	Boerhaavia diffusa	Gajpurna	Nyctaginaceae	Liver enlargement, jaundice, inflammation, antidote to snake bite.	Leaves/ Stem
102	Mirobilis jalapa	Nakta	Nyctaginaceae	Sprain boils, scorpion bite, *etc.*	Root
103	Nyctanthus arbor-tristis	Chirata	Oleaceae	Teeth infections, fever ,*etc.*	Leaves/ Seeds
104	Oxalis corniculata	Chuka buti	Oxalidaceae	Dyspepsia, blood stool,insomnia, snake antidote	Leaves
105	Argemone mexicana	Peli kataieyya	Papaveraceae	Eczema, psoriasis, conjunctivitis, *etc.*	Leaves
106	Piper longum	Pipal	Piperaceae	Abortion, nasal congestion, whooping cough, *etc.*	Fruits
107	Bambusa stricta	Beans	Poaceae	Acidity, asthma, bleeding, diabetes, *etc.*	Leaves

(Table 1) cont.....

S. No.	Botanical Name of the Plant	Local Name	Family	Medicinal Uses	Specific Attributes
108	Zizyphus nummularia	Jhadber	Rhamnaceae	Throat troubles, cough, cold, fever, chicken pox, *etc.*	Leaves/ Stem
109	Xeromphis spinosa	Maini	Rubiaceae	Urinary disorders, such as insecticide, blood purifier, *etc.*	Fruits
110	Aegle marmelos	Bilva	Rutaceae	Diarrhea, dysentery, astringent, diabetes *etc.*	Fruits/ Leaves
111	Solanum torvum	khat-khiri	Solanaceae	Toothache, liver disorders, cough and cold, *etc.*	Fruits
112	Solanum nigrum	Makoiya	Solanaceae	Female disorders, diuretic and narcotic, sexual debility.	Seeds
113	Withania somnifera	Asgandh	Solanaceae	Successful pregnancy, insect bites, immunity booster	Fruits
114	Vitex negundo	Ratalu	Verbenaceae	Cough, cold, fever, diarrhea, arthritis, *etc.*	Leaves/ Flowers
115	Curcuma longa	Haldi	Zingiberaceae	Diarrhea, fever, antiseptic, anthelmintic, and skin problems.	Rhizome
116	Zingiber officinale	Adrak	Zingiberaceae	Cough, cold, arthritis pain, *etc.*	Rhizome
117	Zingiber roseum	Banabyada	Zingiberaceae	Cough, cold, fractured bone, digestive agent, *etc.*	Rhizome

CHAPTER 4

Review of Results

Abstract: This chapter presents findings from a survey focused on the medicinal utilization of 117 plant species spanning 44 families in the Pilibhit Tiger Reserve (PTR) region. Rather than providing concrete results, the information gathered underscores the importance of further research and clinical trials before the application of these plants for medicinal purposes. Caution is advised, and consultation with modern healthcare professionals is recommended to mitigate potential risks. The primary aim of this chapter is to systematically compile diverse survey results, obtained through questionnaires, creating a valuable resource for researchers and herbalists engaged in future investigations, research and development initiatives, and related studies. Within the PTR region, tribal and rural populations employed these plant species for medicinal purposes, distributed across various environments. The breakdown reveals a prevalence of medicinal plants among Herbs [52], Trees [27] species, followed by Shrubs [19] (Table **1**). Families like Apocynaceae, Liliaceae, Asteraceae, Apiaceae, Leeaceae, Acanthaceae, Asclepiadaceae, Amaranthaceae, Euphorbiaceae, are prominently represented with 54 species belonging to these families (Table **2**). The study identifies genera such as Achyranthes, Cassia, Terminalia, Leea, Allium, and Solanum, with some having multiple species within the study area. This chapter serves as a foundational resource for future investigations, emphasizing the importance of rigorous research and clinical validation before the application of these plants for medicinal purposes. It stands as a valuable reference for researchers and herbalists, contributing to the collective knowledge base and guiding future endeavors in the realms of traditional medicine and biodiversity conservation within the PTR region.

Keywords: Biodiversity, Ethnobotanical knowledge, Healing practices, Medicinal plants.

DISCUSSION

The findings of this survey study are more in the form of information rather than as concrete results. They apply these enumerated plants as medicine, which requires clinical trials and further research. Hence, it is always advisable for anyone to consult their doctors for the avoidance of any loss due to the application of these medicinal plants. The chief motive of writing this manuscript is to compile different results based on several questionnaires in a systematic document beneficial for different researchers and herbalists for the future investigations, research and development.

Table 1. The dominant families of medicinal plant in the study region are:

Apocynaceae	09
Liliaceae	07
Asteraceae	06
Apiaceae	06
Leeaceae	06
Acanthaceae	05
Amaranthaceae	05
Euphorbiaceae	05

The study reveals that out of these enumerated herbal plants, 80 were of Ethnopharmacological uses; 13 plants were used in Ethnopharmacology and Ethno-veterinary and 17 in Ethnogynaecology and Ethnopharmacology. At least 7 plants were used in all three: Ethnogynaecology, Ethnopharmacology and Ethnoveterinary (Fig. **1**).

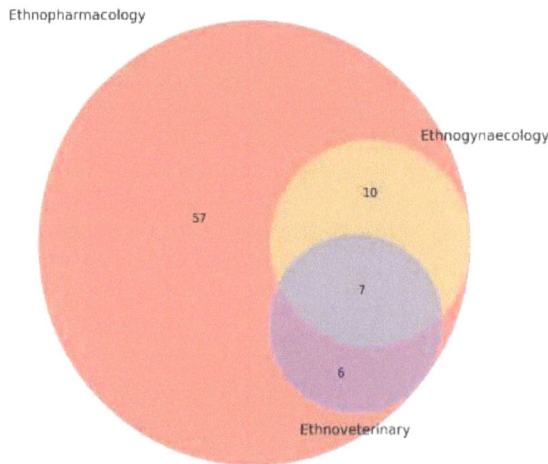

Fig. (1). Diagram of different uses of medicinal plants of the study region.

Ethnic people employed entire plants or suitable plant parts such as root, bulb, tuber, rhizome, stem/wood, leaves (young or adult), stem or root bark, fruits and seeds, or simply seeds only fruit rind, and even flowers. They used varied plant parts as raw crude herbal drug. At the same time, some were prepared to medicine and administered as decoction, infusion, oil, juice, extract, latex, gum, ash, powder, paste, fumes, syrup, *etc* as per the ailment or disease. This may be applied externally, inhaled or consumed orally as per the prescription. In some instances, contact therapy is also significant *e.g.* placing a drug around the neck of a person, in hair, around waist or arm, *etc.* The patients were also, in few cases were advised to consume a drug as a vegetable or it constitutes their diet (*e.g. Momordica charantia*). It is also observed that while preparing medicines, they use honey, sugar, milk, salt, other plant juices, *etc.* In a few instances, they mutter hymns (mantras) also.

The results show that not only routine diseases like cold, cough, fever, diarrhea & dysentery, stomach disorders were cured by administering herbal drugs but tribals have cure for certain chronic diseases like leucoderma (*Abrus precatorious* L., *Hemidesmus indicus* (L.) Willd. & *Vernonia anthelmintica* R.Br., *etc*) leprosy (*Centella asiatica* Urban, *Cassia fistula* L., *Terminalia alata* Heyne, *Albizzia lebbeck* Benth., *etc*), skin diseases, boils and blisters (*Achyranthes aspera* Hook., *Asparagus racemosus* Willd., *Azadirachta indica* A. Juss., *Bombax ceiba* L. *Cassia tora* L., *Mallotus phillipensis* Mueller, *Vitex negundo* L., *etc.*) rheumatism and arthritis (*Achyranthes aspera* Hook., *Litsea glutinosa* (Lour.) C. B. Robins, *Sida rhombifolia* L., *Vitex negundo* L., *Allium sativum* L., *Boerhaavia diffusa* L., *Calotropis gigantean* (L), *Cuscuta reflexa* Roxb., *Peristrophe bicalyculata* Nees, *etc*), asthma (*Achyranthes aspera* var. *porphyristachia* Hook., *Cissampelos pareira* L., *Piper longum* L., *Bambusa stricta* Roxb., *etc*) snake and scorpion bites (*Rauvolfia serpentina* (L.) *Peristrophe bicalyculata* Nees, *Cissampelos pareira* L., *Achyranthes aspera* var *aspera* L., *Mirabilis jalapa* L. *etc*), bone fracture (*Oroxylum indicum* Vent., *Cassia tora* L., *Leea macrophylla* Roxb. and *Peristrophe bicalyculataetc*) and many more. Plants like *Achyranthes asperava. porphyristachia* Hook, *Cassia tora* L., *Nyctanthes arbor-tristis* L., *etc* were used in malarial fever. In contrast, *Alstonia scholars* R. Br., *Azadirachta indica* A. Juss., *Cissampelos pareira* L., *Rauvolfia serpentina* (L.), *Vitex negundo* L., *etc.* were used as febrifuge. *Boerhaavia diffusa* L., *Calotropis gigantean* (L), *Cassia fistula* L., *Cuscuta reflexa* Roxb. *etc.* was used in Jaundice. *Achyranthes aspera* Hook, *Boerhaavia diffusa* L., *Calotropis gigantea* (L), *Cissampelos pareira* L., *Emblica officinalis* Gaertn., *etc.* were used in various ophthalmic troubles. *Achyranthes aspera* Hook, *Boerhaavia diffusa* L., *Cuscuta reflexa* Roxb., *etc.* were used in various venereal diseases. *Aegle marmelos* Correa, *Azadirachta indica* A. Juss., *Syzigium cuminii* L., *Momordica charantia* L. *etc.*, were used to treat diabetes. *Abrus precatorious* L., *Achyranthes aspera* Hook, and *Urginia indica* Kunth were used as antifertility and aborting agents. Some plants are used only externally such as roots of *Achyranthes aspera* var. *porphyristachia* Hook., when tied to the stomach of pregnant women helps in easy delivery. This shows the potency of plants in human health care.

Table 2. The prevalence of medicinal plants in the study region.

Herbs	52
Trees	27
Shrubs	19
Under shrubs	04
Herb Climbers	06
Shrub Climbers	06
Ferns	02
Bamboo	01

The present book indicated that the tribals and ethnic people of the area are conscious of the health of their livestock. They depend very much upon their domestic animals and treat their prevailing ailments like wounds, stomach disorders, eye trouble, fever, bone fracture, *etc.* with medicinal plants' help, Ethnic communities maintain kitchen gardens regularly to meet their dietary and other miscellaneous needs. Some species maintained by them are of medicinal value as well *e.g. Adhatoda zeylanica, Centella asiatica, Trachyspermum ammi, Catharanthu sroseus, Ocimum basilicum, Allium cepa, Allium sativum, Zingiber roseum, Curcuma longa, Momordica charantia etc.* The results satisfy the local populations, so they have great faith in it. However, scientific screening of the recipes is essential for obtaining the active principles and mechanism.

CONCLUSION

In summary, the investigation into the utilization of medicinal plants in the Pilibhit Tiger Reserve (PTR) area reveals a diverse array of biodiversity interwoven with customary therapeutic methodologies. The results, although offering useful insights, underscore the necessity for careful deliberation and more investigation before the extensive implementation of these botanical species for medical use. The importance of seeking guidance from healthcare professionals to minimize potential hazards is emphasized.

The primary objective of this chapter is to methodically gather survey data, resulting in a comprehensive database that can be advantageous for academics and herbalists involved in future investigations, research and development initiatives, and associated scholarly inquiries. The compilation provides insight into the many uses of these enumerated plant species from varied families in the PTR region. It highlights the prevalence of these species among trees and herbs and the significant presence of specific plant families and genera. Prominently emerging in this context are families such as Apocynaceae, Liliaceae, Leeaceae, Apiaceae,

Asteraceae, Acanthaceae, Amaranthaceae, Euphorbiaceae, which serve to underscore their cultural and ecological significance. The discernment of distinct genera encompassing several species contributes further intricacy to comprehending the plant diversity within the location.

This chapter is a fundamental reference, advocating a prudent attitude toward using traditional medicinal knowledge. The study above enhances the overall comprehension of traditional medicine and biodiversity conservation in the PTR region, offering significant insights that can inform future research endeavors. In conclusion, thoroughly examining these discoveries will enhance the sustainable exploitation of botanical resources for therapeutic applications while safeguarding the intricate equilibrium between human welfare and ecological integrity in the PTR area.

REFERENCES

[1]   Palakkal Dse. Reviving India'S Indigenous Knowledge System: An Attempt Towards A 'Sustainable Future. Emerging. 140.

[2]   Lynch JP. Root phenotypes for improved nutrient capture: an underexploited opportunity for global agriculture. New Phytol 2019; 223(2): 548-64.
[http://dx.doi.org/10.1111/nph.15738] [PMID: 30746704]

[3]   Small E. Top 100 food plants. NRC Research Press 2009.

[4]   Udenta EA, Obizoba IC, Oguntibeju OO. Anti-diabetic effects of Nigerian indigenous plant foods/diets. Antioxidant-antidiabetic agents and human health. 2014; 13.

[5]   Rubatzky VE, Yamaguchi M. World vegetables: principles, production, and nutritive values. Springer Science & Business Media 2012.

[6]   Shalchian MM, Arabani M. Application of plant-derived fibers in soil reinforcement on experimental, numerical, and case study scales: a review. Bull Eng Geol Environ 2023; 82(1): 19.
[http://dx.doi.org/10.1007/s10064-022-03029-8]

[7]   Farnsworth NR, Soejarto DD. Global importance of medicinal plants. The conservation of medicinal plants. 1991; 26(26): 25-51.
[http://dx.doi.org/10.1017/CBO9780511753312.005]

[8]   Che C-T, George V, Ijinu T, Pushpangadan P, Andrae-Marobela K. Traditional medicine Pharmacognosy. Elsevier 2017; pp. 15-30.
[http://dx.doi.org/10.1016/B978-0-12-802104-0.00002-0]

[9]   Alamgir A. Therapeutic use of medicinal plants and their extracts. Springer 2017; 1.
[http://dx.doi.org/10.1007/978-3-319-63862-1]

[10]   Muhammad M, Badshah L, Shah AA, *et al.* Ethnobotanical profile of some useful plants and fungi of district Dir Upper, Tehsil Darora, Khyber Pakhtunkhwa, Pakistan. Ethnobot Res Appl 2021; 21: 1-15.
[http://dx.doi.org/10.32859/era.21.42.1-15]

[11]   Karunamoorthi K, Jegajeevanram K, Xavier J, Vijayalakshmi J, Melita L. Tamil traditional medicinal system - siddha: an indigenous health practice in the international perspectives. TANG [HUMANITAS MEDICINE] 2012; 2(2): 12.1-12.11.
[http://dx.doi.org/10.5667/tang.2012.0006]

[12]   Maheshwari A, Singh N, Krishna A, Ramakrishnan G. A benchmark and dataset for post-OCR text correction in Sanskrit. 2022.
[http://dx.doi.org/10.18653/v1/2022.findings-emnlp.466]

[13]   Wujastyk D. Indian manuscripts. Manuscript Cultures: Mapping the Field. 2014: 159-82.
[http://dx.doi.org/10.1515/9783110225631.159]

[14]   Borokini TI, Lawal IO. Traditional medicine practices among the Yoruba people of Nigeria: A historical perspective. Journal of Medicinal Plants Studies 2014; 2(6): 20-33.

[15]   Pisanti S, Bifulco M. Medical *Cannabis* : A plurimillennial history of an evergreen. J Cell Physiol 2019; 234(6): 8342-51.
[http://dx.doi.org/10.1002/jcp.27725] [PMID: 30417354]

[16]   Sharma J, Gairola S, Gaur RD, Painuli RM, Siddiqi TO. Ethnomedicinal plants used for treating epilepsy by indigenous communities of sub-Himalayan region of Uttarakhand, India. J Ethnopharmacol 2013; 150(1): 353-70.
[http://dx.doi.org/10.1016/j.jep.2013.08.052] [PMID: 24029249]

[17]   Dixit G, Vakshasya S. Ethnotherapies of various human ailments by wild plants used by the tharus of Indo-Nepal Sub-Himalayan terai international border region of rohilkhand division of Uttar Pradesh, India. J Tradit Med Clin Natur 2019; 8(279): 2.

[18]   Mishra A, Shrivastava P. Ethnobotanical study of tharu community of Panchapedwa, Balrampur, Uttar Pradesh. J Pharmacogn Phytochem 2018; 7(1): 1626-8.

[19]   AGN F, Modjenpa N, Veranso M. Ethnobotany of acanthaceae in the Mount Cameroon region. 2013: 2859-2866.

[20]   Gentry AH. Changes in plant community diversity and floristic composition on environmental and geographical gradients. Ann Mo Bot Gard 1988; 75(1): 1-34.
        [http://dx.doi.org/10.2307/2399464]

[21]   Layek U, Midday M, Bisui S, Kundu A, Karmakar P. Floral biology, breeding system and pollination ecology of *Justicia betonica* L. (Acanthaceae): An assessment of its low reproductive success in West Bengal, India. Plant Species Biol 2022; 37(4): 278-93.
        [http://dx.doi.org/10.1111/1442-1984.12380]

[22]   Gangwar AK, Ghosh AK. Medicinal uses and pharmacological activity of Adhatoda vasica. Int J Herb Med 2014; 2(1): 88-91.

[23]   Niu Z, Lin Z, Tong Y, Chen X, Deng Y. Complete plastid genome structure of thirteen Asian Justicia (Acanthaceae) species: comparative genomics and phylogenetic analyses. 2023.

[24]   Robertson KR. The genera of Amaranthaceae in the southeastern United States. J Arnold Arbor 1981; 62(3): 267-313.
        [http://dx.doi.org/10.5962/bhl.part.11251]

[25]   Thorat BR. Review on Celosia argentea L. Plant Research Journal of Pharmacognosy and Phytochemistry 2018; 10(1): 109-19.

[26]   Jan HA, Jan S, Bussmann RW, Ahmad L, Wali S, Ahmad N. Ethnomedicinal survey of the plants used for gynecological disorders by the indigenous community of district Buner, Pakistan. Ethnobot Res Appl 2020; 19: 1-18.
        [http://dx.doi.org/10.32859/era.19.26.1-18]

[27]   McGaw LJ, Eloff JN. Ethnoveterinary use of southern African plants and scientific evaluation of their medicinal properties. J Ethnopharmacol 2008; 119(3): 559-74.
        [http://dx.doi.org/10.1016/j.jep.2008.06.013] [PMID: 18620038]

[28]   Cock IE, Selesho MI, Van Vuuren SF. A review of the traditional use of southern African medicinal plants for the treatment of selected parasite infections affecting humans. J Ethnopharmacol 2018; 220: 250-64.
        [http://dx.doi.org/10.1016/j.jep.2018.04.001] [PMID: 29621583]

[29]   Pell SK, Mitchell J, Miller A, Lobova T. Anacardiaceae Flowering plants Eudicots: Sapindales. Cucurbitales, Myrtaceae 2011; pp. 7-50.

[30]   Mitchell JD. The poisonous Anacardiaceae genera of the world. Advances in Economic Botany 1990; pp. 103-29.

[31]   Sonowal R, Barua I. Ethnomedical practices among the Tai-Khamyangs of Assam, India. Stud Ethno-Med 2011; 5(1): 41-50.
        [http://dx.doi.org/10.1080/09735070.2011.11886390]

[32]   Titus B. A comparative study on the health management practices and the use of ethno-veterinary medicine in the treatment of opr in goats and sheep in karamoja region: Uganda Martyrs University; 2022.

[33]   Nowak A, Nobis M, Nowak S, *et al.* Illustrated flora of Tajikistan and adjacent areas. Warsaw: Polish Academy of Sciences. Botanical Garden Center for Biological 2020.

[34] Wang XJ, Luo Q, Li T, *et al.* Origin, evolution, breeding, and omics of Apiaceae: a family of vegetables and medicinal plants. Hortic Res 2022; 9: uhac076.
[http://dx.doi.org/10.1093/hr/uhac076] [PMID: 38239769]

[35] Downie SR. JANSEN RMPRK, Downie SR, Peery RM, Jansen RK. Another first for the Apiaceae: evidence for mitochondrial DNA transfer into the plastid genome. Journal of Faculty of Pharmacy of Istanbul University 2015; 44(2): 131-44.

[36] El-Ahmady S, Ibrahim N, Farag N, Gabr S. Apiaceae plants growing in the east. Ethnopharmacology of Wild Plants. 2021.
[http://dx.doi.org/10.1201/9781003052814-15]

[37] Gutiérrez-Grijalva EP, López-Martínez LX, Contreras-Angulo LA, Elizalde-Romero CA, Heredia JB. Plant alkaloids: Structures and bioactive properties. Plant-Derived Bioactives: Chemistry and Mode of Action. 2020: 85-117.

[38] Endress ME, Meve U, Middleton DJ, Liede-Schumann S. Flowering Plants Eudicots: Apiales. Gentianales 1905; 2018: 207-411. [except Rubiaceae].

[39] Islam MS, Lucky RA. A study on different plants of Apocynaceae family and their medicinal uses. J Pharm Res 2019; 4(1): 40-4.

[40] Bhadane BS, Patil MP, Maheshwari VL, Patil RH. Ethnopharmacology, phytochemistry, and biotechnological advances of family Apocynaceae: A review. Phytother Res 2018; 32(7): 1181-210.
[http://dx.doi.org/10.1002/ptr.6066] [PMID: 29575195]

[41] Wong SK, Lim YY, Chan EW. Botany, uses, phytochemistry and pharmacology of selected Apocynaceae species: A review. Pharmacogn Commun 2013; 3(3).

[42] Ollerton J, Liede-Schumann S, Endress ME, *et al.* The diversity and evolution of pollination systems in large plant clades: Apocynaceae as a case study. Ann Bot (Lond) 2019; 123(2): 311-25.
[http://dx.doi.org/10.1093/aob/mcy127] [PMID: 30099492]

[43] Wanntorp L, Grudinski M, Forster PI, Muellner-Riehl AN, Grimm GW. Wax plants (*Hoya*, Apocynaceae) evolution: Epiphytism drives successful radiation. Taxon 2014; 63(1): 89-102.
[http://dx.doi.org/10.12705/631.3]

[44] Nabout JC, Magalhães MR, de Amorim Gomes MA, da Cunha HF. The impact of global climate change on the geographic distribution and sustainable harvest of Hancornia speciosa Gomes (Apocynaceae) in Brazil. Environ Manage 2016; 57(4): 814-21.
[http://dx.doi.org/10.1007/s00267-016-0659-5] [PMID: 26796699]

[45] Bayton R, Maughan S. Plant families: a guide for gardeners and botanists. University of Chicago Press 2017.
[http://dx.doi.org/10.7208/chicago/9780226536675.001.0001]

[46] Ivancic A, Lebot V. Botany and genetics of New Caledonian wild taro, Colocasia esculenta. 1999.

[47] Rahayu S, Hakim L, Indriyani S, Sukenti K, Eds. Ethnobotany and conservation of Araceae of Sasak community in Ende, Sengkol Village, Central Lombok IOP Conference Series: Earth and Environmental Science. IOP Publishing 2022.

[48] Motley TJ. The ethnobotany of sweet flag,acorus Calamus (Araceae). Econ Bot 1994; 48(4): 397-412.
[http://dx.doi.org/10.1007/BF02862235]

[49] Gurib-Fakim A. Medicinal plants: Traditions of yesterday and drugs of tomorrow. Mol Aspects Med 2006; 27(1): 1-93.
[http://dx.doi.org/10.1016/j.mam.2005.07.008] [PMID: 16105678]

[50] Listiani L, Abrori FM. Ethnobotanical study on Tidung Tribe in using plants for medicine, spice, and ceremony. IPTEK The Journal for Technology and Science 2019; 29(1): 18-24.
[http://dx.doi.org/10.12962/j20882033.v29i1.3057]

[51] Rahman MA, Wilcock CC. Diversity of life-form and distribution of the Asclepiadaceae in south-west Asia and the Indian subcontinent. J Biogeogr 1992; 19(1): 51-8.
      [http://dx.doi.org/10.2307/2845619]

[52] Ratageri RH, Gangadhar G. Phytodiversity of Asclepiadaceae in Jogimatti. Chitradurga, Karnataka: Applied Aquatic and Terrestrial Eco-Biology 2022; p. 163.

[53] Reddy SH, Chakravarthi M, Chandrashekara K, Naidu C. Phytochemical screening and antibacterial studies on leaf and root extracts of asclepias curassavica (L). Journal of Pharmacy and Biological Sciences. 2012: 39-44.

[54] Al-Qura'n S. Ethnopharmacological survey of wild medicinal plants in Showbak, Jordan. J Ethnopharmacol 2009; 123(1): 45-50.
      [http://dx.doi.org/10.1016/j.jep.2009.02.031] [PMID: 19429338]

[55] Rajakumar N, Shivanna MB. Ethno-medicinal application of plants in the eastern region of Shimoga district, Karnataka, India. J Ethnopharmacol 2009; 126(1): 64-73.
      [http://dx.doi.org/10.1016/j.jep.2009.08.010] [PMID: 19686831]

[56] Ninich O, Et-tahir A, Kettani K, *et al.* Plant sources, techniques of production and uses of tar: A review. J Ethnopharmacol 2022; 285: 114889.
      [http://dx.doi.org/10.1016/j.jep.2021.114889] [PMID: 34864129]

[57] Sharma LK, Dadhich N, Kumar A. Plant based veterinary medicine from traditional knowledge of India. Nelumbo 2005; 47(1-4): 43-52.

[58] Snively-Martinez AE. Family poultry systems on the southern pacific coast of Guatemala: Livelihoods, ethnoveterinary medicine and healthcare decision making. Washington State University 2017.

[59] Sharma SK, Alam A. Biological prospecting of the "Hidden Diversity" of medicinal plants (Asteraceae) in south-eastern Rajasthan. 2022.

[60] Bottoni M, Milani F, Colombo L, *et al.* Using medicinal plants in Valmalenco (Italian Alps): From tradition to scientific approaches. Molecules 2020; 25(18): 4144.
      [http://dx.doi.org/10.3390/molecules25184144] [PMID: 32927742]

[61] Lara Reimers EA, Fernández Cusimamani E, Lara Rodríguez EA, Zepeda del Valle JM, Polesny Z, Pawera L. An ethnobotanical study of medicinal plants used in Zacatecas state, Mexico. Acta Soc Bot Pol 2018; 87(2).
      [http://dx.doi.org/10.5586/asbp.3581]

[62] Dajue L, Mündel H-H. Safflower, Carthamus tinctorius L: Bioversity International; 1996.

[63] Bussmann RW, Sharon D. Traditional medicinal plant use in Northern Peru: tracking two thousand years of healing culture. J Ethnobiol Ethnomed 2006; 2(1): 47.
      [http://dx.doi.org/10.1186/1746-4269-2-47] [PMID: 17090303]

[64] Güler O, Polat R, Karaköse M, Çakılcıoğlu U, Akbulut S. An ethnoveterinary study on plants used for the treatment of livestock diseases in the province of Giresun (Turkey). S Afr J Bot 2021; 142: 53-62.
      [http://dx.doi.org/10.1016/j.sajb.2021.06.003]

[65] Gasson P, Dobbins DR. Wood anatomy of the Bignoniaceae, with a comparison of trees and lianas. IAWA J 1991; 12(4): 389-415.
      [http://dx.doi.org/10.1163/22941932-90000541]

[66] Gentry AH. Bignoniaceae: part I (Crescentieae and tourrettieae). Flora Neotropica 1980; 25(1): 1-130.

[67] Zhang J, Hunto ST, Yang Y, Lee J, Cho JY. Tabebuia impetiginosa: a comprehensive review on traditional uses, phytochemistry, and immunopharmacological properties. Molecules 2020; 25(18): 4294.
      [http://dx.doi.org/10.3390/molecules25184294] [PMID: 32962180]

[68]   Nesheim I, Dhillion SS, Anne Stølen K. What happens to traditional knowledge and use of natural resources when people migrate? Hum Ecol Interdiscip J 2006; 34(1): 99-131.
[http://dx.doi.org/10.1007/s10745-005-9004-y]

[69]   Kathambi V, Mutie FM, Rono PC, *et al.* Traditional knowledge, use and conservation of plants by the communities of Tharaka-Nithi County, Kenya. Plant Divers 2020; 42(6): 479-87.
[http://dx.doi.org/10.1016/j.pld.2020.12.004] [PMID: 33733015]

[70]   Batistini AP, Telles MPdC, Bertoni BW, *et al.*, Genetic diversity of natural populations of Anemopaegma arvense (Bignoniaceae) in the Cerrado of São Paulo State, Brazil. 2009.

[71]   Idris A, Al-tahir I, Idris E. Antibacterial activity of endophytic fungi extracts from the medicinal plant Kigelia africana. Egypt Acad J Biol Sci G Microbiol 2013; 5(1): 1-9.
[http://dx.doi.org/10.21608/eajbsg.2013.16639]

[72]   Refaat JY, Desoukey SA, Ramadan MS, Kamel M. Bombacaceae between the ethnomedical uses and pharmacological evidences: A review. Nat Prod J 2014; 4(4): 254-70.
[http://dx.doi.org/10.2174/2210315504666141125003412]

[73]   Sidibe M, Williams JT. Baobab, Adansonia Digitata L: Crops for the Future; 2002.

[74]   Rashford J. The use of baobab leaves (Adansonia digitata L.) for food in Africa: a review. Econ Bot 2018; 72(4): 478-95.
[http://dx.doi.org/10.1007/s12231-018-9438-y]

[75]   Baum DA. A Systematic Revision of Adansonia (Bombacaceae). Ann Mo Bot Gard 1995; 82(3): 440-71.
[http://dx.doi.org/10.2307/2399893]

[76]   Chaturvedi N, Singhal S. An ethnobotanical study of medicinal plants in jaipur district and adjoining area. Ann Rom Soc Cell Biol 2020; 2233-46.

[77]   Masola S, Mosha R, Wambura P. Assessment of antimicrobial activity of crude extracts of stem and root barks from Adansonia digitata (Bombacaceae)(African baobab). Afr J Biotechnol 2009; 8(19).

[78]   Bhatt P, Pandya KB, Patel U, Patel H, Modi C. Survey on ethnoveterinary practices around junagadh, Gujarat, India. Indian J Pharm Sci 2019; 81(1).
[http://dx.doi.org/10.4172/pharmaceutical-sciences.1000493]

[79]   Meudt HM, Prebble JM, Lehnebach CA. Native New Zealand forget-me-nots (Myosotis, Boraginaceae) comprise a Pleistocene species radiation with very low genetic divergence. Plant Syst Evol 2015; 301(5): 1455-71.
[http://dx.doi.org/10.1007/s00606-014-1166-x]

[80]   Wollenweber E, Wehde R, Dörr M, Stevens JF. On the occurrence of exudate flavonoids in the borage family (Boraginaceae). Z Naturforsch C J Biosci 2002; 57(5-6): 445-8.
[http://dx.doi.org/10.1515/znc-2002-5-607] [PMID: 12132682]

[81]   Riedl H. Boraginaceae. Flora Malesiana-Series 1. Spermatophyta 1997; 13(1): 43-144.

[82]   Nyamwamu N, Okari O, Gisesa W. A survey of medicinal plants used by the gusii community in the treatment of digestive disorders and other inflammatory conditions. Faslnamah-i Giyahan-i Daruyi 2020; 8: 21-33.

[83]   Ashagre M, Asfaw Z, Kelbessa E. Ethnobotanical study of wild edible plants in Burji District, Segan area zone of southern nations, nationalities and peoples region (SNNPR), Ethiopia. J Ethnobiol Ethnomed 2016; 12(1): 32.
[http://dx.doi.org/10.1186/s13002-016-0103-1] [PMID: 27485265]

[84]   Giday M, Asfaw Z, Elmqvist T, Woldu Z. An ethnobotanical study of medicinal plants used by the Zay people in Ethiopia. J Ethnopharmacol 2003; 85(1): 43-52.
[http://dx.doi.org/10.1016/S0378-8741(02)00359-8] [PMID: 12576201]

[85]   Pirbalouti AG, Yousefi M, Nazari H, Karimi I, Koohpayeh A. Evaluation of burn healing properties of Arnebia euchroma and Malva sylvestris. Electronic Journal of Biology 2009; 5(3): 62-6.

[86]   Regassa R. Assessment of indigenous knowledge of medicinal plant practice and mode of service delivery in Hawassa city, southern Ethiopia. J Med Plants Res 2013; 7(9): 517-35.

[87]   Palmer WA, Pullen KR. The phytophagous arthropods associated with Senna obtusifolia (Caesalpiniaceae) in Mexico and Honduras and their prospects for utilization for biological control. Biol Control 2001; 20(1): 76-83.
[http://dx.doi.org/10.1006/bcon.2000.0879]

[88]   Seethapathy GS, Ganesh D, Santhosh Kumar JU, *et al.* Assessing product adulteration in natural health products for laxative yielding plants, Cassia, Senna, and Chamaecrista, in Southern India using DNA barcoding. Int J Legal Med 2015; 129(4): 693-700.
[http://dx.doi.org/10.1007/s00414-014-1120-z] [PMID: 25425095]

[89]   Reddy SH, Al-Kalbani AS, Al-Rawahi AS. Studies on phytochemical screening-GC-MS characterization, antimicrobial and antioxidant assay of black cumin seeds (nigella sativa) and senna alexandria (cassia angustifolia) solvent extracts. Int J Pharm Sci Res 2018; 9(2): 490-7.

[90]   Sale JB. The importance and values of wild plants and animals in Africa: Iucn; 1983.

[91]   Abdelrahman GH, Mariod AA. Tamarindus indica: phytochemical constituents, bioactive compounds and traditional and medicinal uses Wild Fruits: Composition. Nutritional Value and Products 2019; pp. 229-38.

[92]   Kiruba S, Jeeva S, Dhas S. Enumeration of ethnoveterinary plants of cape comorin. Tamil Nadu 2006.

[93]   Mahanal S., Zubaidah S., Julung H., Ege B. Ethnobotany of traditional medicinal plants used by Dayak desa Community in Sintang, West Kalimantan, Indonesia. Biodiversitas (Surak) 2019; 20(5).

[94]   Ouvrard D, Chalise P, Percy DM. Host-plant leaps versus host-plant shuffle: a global survey reveals contrasting patterns in an oligophagous insect group (Hemiptera, Psylloidea). Syst Biodivers 2015; 13(5): 434-54.
[http://dx.doi.org/10.1080/14772000.2015.1046969]

[95]   Jacobs M. Capparidaceae. Flora Malesiana-Series 1. Spermatophyta 1960; 6(1): 61-105.

[96]   Al-Zubaidy AA, Khalil AM. Gastroprotective effect of capparis spinosa on indomethacin-induced gastric ulcer in rats. Arch Razi Inst 2022; 77(4): 1429-37.
[PMID: 36883165]

[97]   Rivera D, Inocencio C, Obón C, Alcaraz F. Review of food and medicinal uses ofCapparis L. SubgenusCapparis (capparidaceae). Econ Bot 2003; 57(4): 515-34.
[http://dx.doi.org/10.1663/0013-0001(2003)057[0515:ROFAMU]2.0.CO;2]

[98]   Moghaddasi MS. Caper (Capparis spp.) importance and medicinal usage. Adv Environ Biol 2011; 872-80.

[99]   Orwa JA, Jondiko IJO, Minja RJA, Bekunda M. The use of Toddalia asiatica (L) Lam. (Rutaceae) in traditional medicine practice in East Africa. J Ethnopharmacol 2008; 115(2): 257-62.
[http://dx.doi.org/10.1016/j.jep.2007.09.024] [PMID: 17996412]

[100]   Janz N, Nylin S. Butterflies and plants: a phylogenetic study. Evolution 1998; 52(2): 486-502.
[http://dx.doi.org/10.2307/2411084] [PMID: 28568350]

[101]   De Morais Lima GR, De Sales IRP, Caldas Filho MRD, *et al.* Bioactivities of the genus Combretum (Combretaceae): a review. Molecules 2012; 17(8): 9142-206.
[http://dx.doi.org/10.3390/molecules17089142] [PMID: 22858840]

[102]   Begum T. Comparative study on Combretum and Terminalia species of the Combretaceae family. Communications 2002; 1(1).

[103]   Mongalo NI, McGaw LJ, Segapelo TV, Finnie JF, Van Staden J. Ethnobotany, phytochemistry,

toxicology and pharmacological properties of Terminalia sericea Burch. ex DC. (Combretaceae) – A review. J Ethnopharmacol 2016; 194: 789-802.
[http://dx.doi.org/10.1016/j.jep.2016.10.072] [PMID: 27989875]

[104]   Chomicki G, Schaefer H, Renner SS. Origin and domestication of Cucurbitaceae crops: insights from phylogenies, genomics and archaeology. New Phytol 2020; 226(5): 1240-55.
[http://dx.doi.org/10.1111/nph.16015] [PMID: 31230355]

[105]   White JC. Differential bioavailability of field-weathered p,p′-DDE to plants of the Cucurbita and Cucumis genera. Chemosphere 2002; 49(2): 143-52.
[http://dx.doi.org/10.1016/S0045-6535(02)00277-1] [PMID: 12375861]

[106]   Hayatu M., Mustapha Y., Sani L.A. Nutritional and anti-nutritional properties of the seeds of six selected Nigerian Cucurbit Germplasm. J Plant Dev 2021; 28: 139-50.

[107]   Basch E, Gabardi S, Ulbricht C. Bitter melon (Momordica charantia): A review of efficacy and safety. Am J Health Syst Pharm 2003; 60(4): 356-9.
[http://dx.doi.org/10.1093/ajhp/60.4.356] [PMID: 12625217]

[108]   Akoroda MO. Ethnobotany ofTelfairia occidentalis (cucurbitaceae) among Igbos of Nigeria. Econ Bot 1990; 44(1): 29-39.
[http://dx.doi.org/10.1007/BF02861064]

[109]   Bekele D, Asfaw Z, Petros B, Tekie H. Ethnobotanical study of plants used for protection against insect bite and for the treatment of livestock health problems in rural areas of Akaki District, Eastern Shewa, Ethiopia. Topclass Journal of Herbal Medicine 2012; 1(2): 12-24.

[110]   Costea M, García MA, Stefanović S. A phylogenetically based infrageneric classification of the parasitic plant genus Cuscuta (dodders, Convolvulaceae). Syst Bot 2015; 40(1): 269-85.
[http://dx.doi.org/10.1600/036364415X686567]

[111]   Bylina M, Kolokolova NM , Eds. Dodder and How to deal with it. pp 110-114. XI International Scientific and Practical conference, Astrakhan University, Russia

[112]   Nickrent DL. Parasitic angiosperms: How often and how many? Taxon 2020; 69(1): 5-27.
[http://dx.doi.org/10.1002/tax.12195]

[113]   Musselman L, Press M, Press M, Graves J. Parasitic plants. Parasitic plants. 1995: 1.

[114]   Saeed Khattak N, Nouroz F, Ur Rahman I, Noreen S. Ethno veterinary uses of medicinal plants of district Karak, Pakistan. J Ethnopharmacol 2015; 171: 273-9.
[http://dx.doi.org/10.1016/j.jep.2015.05.048] [PMID: 26055343]

[115]   Bungard RA. Photosynthetic evolution in parasitic plants: insight from the chloroplast genome. BioEssays 2004; 26(3): 235-47.
[http://dx.doi.org/10.1002/bies.10405] [PMID: 14988925]

[116]   Kumar S, Das G, Shin HS, Patra JK. Dioscorea spp.(a wild edible tuber): a study on its ethnopharmacological potential and traditional use by the local people of Similipal Biosphere Reserve, India. Front Pharmacol 2017; 8: 52.
[http://dx.doi.org/10.3389/fphar.2017.00052] [PMID: 28261094]

[117]   Luo GF, Podolyan A, Kidanemariam DB, Pilotti C, Houliston G, Sukal AC. A review of viruses infecting yam (Dioscorea spp.). Viruses 2022; 14(4): 662.
[http://dx.doi.org/10.3390/v14040662] [PMID: 35458392]

[118]   Chiu CS, Deng JS, Chang HY, *et al.* Antioxidant and anti-inflammatory properties of taiwanese yam (Dioscorea japonica Thunb. var. pseudojaponica (Hayata) Yamam.) and its reference compounds. Food Chem 2013; 141(2): 1087-96.
[http://dx.doi.org/10.1016/j.foodchem.2013.04.031] [PMID: 23790890]

[119]   Okeke EC, Eneobong HN, Uzuegbunam AO, Ozioko AO, Kuhnlein H. Igbo traditional food system: Documentation, uses and research needs. Pak J Nutr 2008; 7(2): 365-76.

[http://dx.doi.org/10.3923/pjn.2008.365.376]

[120] Dhakal A, Khanal S, Pandey M. Ethnoveterinary practice of medicinal plants in Chhatradev Rural Municipality, Arghakhanchi District of Western Nepal. Nusantara Bioscience. 2021; 13(1).

[121] Bruchac M. Indigenous knowledge and traditional knowledge. 2014.
[http://dx.doi.org/10.1007/978-1-4419-0465-2_10]

[122] Andriamparany JN, Brinkmann K, Jeannoda V, Buerkert A. Effects of socio-economic household characteristics on traditional knowledge and usage of wild yams and medicinal plants in the Mahafaly region of south-western Madagascar. J Ethnobiol Ethnomed 2014; 10(1): 82.
[http://dx.doi.org/10.1186/1746-4269-10-82] [PMID: 25551198]

[123] Vasas A, Hohmann J. Euphorbia diterpenes: isolation, structure, biological activity, and synthesis (2008-2012). Chem Rev 2014; 114(17): 8579-612.
[http://dx.doi.org/10.1021/cr400541j] [PMID: 25036812]

[124] Wee YC, Gopalakrishnakone P. A colour guide to dangerous plants. NUS Press 1990.

[125] Bhattacharjee A, Bhowmik M, Paul C, Das Chowdhury B, Debnath B. Rubber tree seed utilization for green energy, revenue generation and sustainable development– A comprehensive review. Ind Crops Prod 2021; 174: 114186.
[http://dx.doi.org/10.1016/j.indcrop.2021.114186]

[126] Kichu M, Malewska T, Akter K, *et al.* An ethnobotanical study of medicinal plants of Chungtia village, Nagaland, India. J Ethnopharmacol 2015; 166: 5-17.
[http://dx.doi.org/10.1016/j.jep.2015.02.053] [PMID: 25747148]

[127] Nair B, Punniamurthy N, Kumar S. Ethno-veterinary practices for animal health and the associated Medicinal Plants from 24 Locations in 10 States of India. Res J Vet Sci 2017; 3: 16-25.

[128] Hooli LJ, Jauhiainen JS. Building an innovation system and indigenous knowledge in Namibia. Afr J Sci Technol Innov Dev 2018; 10(2): 183-96.
[http://dx.doi.org/10.1080/20421338.2018.1436737]

[129] Stebbins GL. Polyploidy, hybridization, and the invasion of new habitats. Ann Mo Bot Gard 1985; 72(4): 824-32.
[http://dx.doi.org/10.2307/2399224]

[130] Kargozar R, Azizi H, Salari R. A review of effective herbal medicines in controlling menopausal symptoms. Electron Physician 2017; 9(11): 5826-33.
[http://dx.doi.org/10.19082/5826] [PMID: 29403626]

[131] Hughes J, Pearson E, Grafenauer S. Legumes—a comprehensive exploration of global food-based dietary guidelines and consumption. Nutrients 2022; 14(15): 3080.
[http://dx.doi.org/10.3390/nu14153080] [PMID: 35956258]

[132] Chepape R, Mbatha K, Luseba D. Local use and knowledge validation of fodder trees and shrubs browsed by livestock in Bushbuckridge area, South Africa. Ghanaian Popul 2014; 77: 20-47.

[133] Ali-Shtayeh MS, Jamous RM, Al-Shafie' JH, *et al.* Traditional knowledge of wild edible plants used in Palestine (Northern West Bank): A comparative study. J Ethnobiol Ethnomed 2008; 4(1): 13.
[http://dx.doi.org/10.1186/1746-4269-4-13] [PMID: 18474107]

[134] Hamed AN, Attia E, Desoukey SY. A review on various classes of secondary metabolites and biological activities of Lamiaceae (Labiatae)(2002-2018). Journal of Advanced Biomedical and Pharmaceutical Sciences 2021; 4(1): 16-31.

[135] Frezza C, Venditti A, Serafini M, Bianco A. Phytochemistry, chemotaxonomy, ethnopharmacology, and nutraceutics of Lamiaceae. Studies in natural products chemistry. 2019; 62: 125-78.
[http://dx.doi.org/10.1016/B978-0-444-64185-4.00004-6]

[136] Hassanzadeh MK, Emami SA, Asili J, Najaran ZT. Review of the essential oil composition of Iranian Lamiaceae+. J Essent Oil Res 2011; 23(1): 35-74.

[http://dx.doi.org/10.1080/10412905.2011.9700429]

[137] McKay DL, Blumberg JB. A review of the bioactivity and potential health benefits of peppermint tea (*Mentha piperita* L.). Phytother Res 2006; 20(8): 619-33.
[http://dx.doi.org/10.1002/ptr.1936] [PMID: 16767798]

[138] Huang X, Meyers PA, Wu W, Jia C, Xie S. Significance of long chain iso and anteiso monomethyl alkanes in the Lamiaceae (mint family). Org Geochem 2011; 42(2): 156-65.
[http://dx.doi.org/10.1016/j.orggeochem.2010.11.008]

[139] Bullitta S, Re GA, Manunta MDI, Piluzza G. Traditional knowledge about plant, animal, and mineral-based remedies to treat cattle, pigs, horses, and other domestic animals in the Mediterranean island of Sardinia. J Ethnobiol Ethnomed 2018; 14(1): 50.
[http://dx.doi.org/10.1186/s13002-018-0250-7] [PMID: 30029686]

[140] Ivanova T, Bosseva Y, Chervenkov M, Dimitrova D. Lamiaceae plants in Bulgarian rural livelihoods—Diversity, utilization, and traditional knowledge. Agronomy (Basel) 2022; 12(7): 1631.
[http://dx.doi.org/10.3390/agronomy12071631]

[141] Chaw SM, Liu YC, Wu YW, *et al.* Stout camphor tree genome fills gaps in understanding of flowering plant genome evolution. Nat Plants 2019; 5(1): 63-73.
[http://dx.doi.org/10.1038/s41477-018-0337-0] [PMID: 30626928]

[142] Schmidt BM, Cheng DMK. Ethnobotany: A phytochemical perspective. John Wiley & Sons 2017.
[http://dx.doi.org/10.1002/9781118961933]

[143] Schroeder C. Some useful plants of the botanical family Lauraceae. Calif Avocado Soc Yearb 1976; 59: 30-4.

[144] Debjit Bhowmik C, Kumar K, Chandira M, Jayakar B. Turmeric: a herbal and traditional medicine. Arch Appl Sci Res 2009; 1(2): 86-108.

[145] Balick MJ, Cox PA. Plants, people, and culture: the science of ethnobotany: Garland Science; 2020.

[146] Wang J, Seyler BC, Ticktin T, Zeng Y, Ayu K. An ethnobotanical survey of wild edible plants used by the Yi people of Liangshan Prefecture, Sichuan Province, China. J Ethnobiol Ethnomed 2020; 16(1): 10.
[http://dx.doi.org/10.1186/s13002-019-0349-5] [PMID: 32102675]

[147] Suwardi AB, Syamsuardi . Ethnobotany and conservation of indigenous edible fruit plants in South Aceh, Indonesia. Biodiversitas (Surak) 2020; 21(5).
[http://dx.doi.org/10.13057/biodiv/d210511]

[148] Tillekaratne K, Edirisinghe JP, Gunatilleke CVS, Karunaratne WAIP. Survey of thrips in Sri Lanka: A checklist of thrips species, their distribution and host plants. Ceylon J Sci Biol Sci 2012; 40(2): 89-108.
[http://dx.doi.org/10.4038/cjsbs.v40i2.3926]

[149] Swarnkar S, Katewa S. Ethnobotanical observation on tuberous plants from tribal area of Rajasthan (India). Ethnobotanical leaflets. 2008; 2008(1): 87.

[150] Dapar MLG, Alejandro GJD, Meve U, Liede-Schumann S. Quantitative ethnopharmacological documentation and molecular confirmation of medicinal plants used by the Manobo tribe of Agusan del Sur, Philippines. J Ethnobiol Ethnomed 2020; 16(1): 14.
[http://dx.doi.org/10.1186/s13002-020-00363-7] [PMID: 32138749]

[151] Ndenecho EN. Ethnobotanic resources of tropical montane forests: Indigenous uses of plants in the Cameroon Highland Ecoregion: African Books Collective; 2011.

[152] Abe R, Ohtani K. An ethnobotanical study of medicinal plants and traditional therapies on Batan Island, the Philippines. J Ethnopharmacol 2013; 145(2): 554-65.
[http://dx.doi.org/10.1016/j.jep.2012.11.029] [PMID: 23183086]

[153] Rosengarten F Jr. The book of edible nuts. Courier Corporation 2004.

[154] Zuidema PA. Ecology and management of the Brazil nut tree (Bertholletia excelsa). Promab 2003.

[155] Mori SA, Prance GT. Taxonomy, ecology, and economic botany of the Brazil nut (Bertholletia excelsa Humb & Bonpl: Lecythidaceae). Advances in Economic Botany 1990; pp. 130-50.

[156] Bradacs G, Heilmann J, Weckerle CS. Medicinal plant use in Vanuatu: A comparative ethnobotanical study of three islands. J Ethnopharmacol 2011; 137(1): 434-48.
[http://dx.doi.org/10.1016/j.jep.2011.05.050] [PMID: 21679762]

[157] Mallik BK, Panda T, Padhy RN. Ethnoveterinary practices of aborigine tribes in Odisha, India. Asian Pac J Trop Biomed 2012; 2(3): S1520-5.
[http://dx.doi.org/10.1016/S2221-1691(12)60447-X]

[158] Olounladé PA, Dansou CC, Songbé O, *et al.* Zootechnical, pharmacological uses and chemical composition of napoleonaea vogelii hook & planch (Lecythidaceae) in West Africa—A review. Am J Plant Sci 2021; 12(8): 1288-303.
[http://dx.doi.org/10.4236/ajps.2021.128090]

[159] Shrestha K, Bhattarai S, Bhandari P. Handbook of flowering plants of nepal (Vol. 1 gymnosperms and angiosperms: cycadaceae-betulaceae): Scientific publishers; 2018.

[160] Khalfaoui A, Bouheroum M. Isolation, structural determination of secondary metabolites and biological activities of asphodelus tenuifolius (Liliaceae): Université Frères Mentouri-Constantine 1; 2018.

[161] Pałka P, Cioć M, Hura K, Szewczyk-Taranek B, Pawłowska B. Adventitious organogenesis and phytochemical composition of Madonna lily (Lilium candidum L.) in vitro modeled by different light quality. Plant Cell Tissue Organ Cult 2023; 152(1): 99-114. [PCTOC].
[http://dx.doi.org/10.1007/s11240-022-02391-5]

[162] Zaccai M, Yarmolinsky L, Khalfin B, *et al.* Medicinal properties of Lilium candidum L. and its phytochemicals. Plants 2020; 9(8): 959.
[http://dx.doi.org/10.3390/plants9080959] [PMID: 32751398]

[163] Manoranjotham M, Kamaraj M. Ethnoveterinary usage of medicinal plants in pachamalai hills, Tamil Nadu, India. Journal of Applied and Advanced Research 2016; 1(3): 31-6.
[http://dx.doi.org/10.21839/jaar.2016.v1i3.31]

[164] Palit D, Banerjee A. Traditional uses and conservative lifestyle of lepcha tribe through sustainable bioresource utilization-case studies from darjeeling and north sikkim, INDIA. Int J Conserv Sci 2016; 7(3).

[165] Anyanwu GO, Nisar-ur-Rehman , Onyeneke CE, Rauf K. Medicinal plants of the genus Anthocleista—A review of their ethnobotany, phytochemistry and pharmacology. J Ethnopharmacol 2015; 175: 648-67.
[http://dx.doi.org/10.1016/j.jep.2015.09.032] [PMID: 26432351]

[166] Behera MC, Mohanty TL, Paramanik BK. Silvics, phytochemistry and ethnopharmacy of endangered poison nut tree (Strychnos nux-vomica L.): A review. J Pharmacogn Phytochem 2017; 6(5): 1207-16.

[167] Behera MC. Strychnos nux-vomica Linn-Strychnine Tree. 2019; 2: 390-404.

[168] Deeba F, Muhammad G, Iqbal Z, Hussain I. Appraisal of ethno-veterinary practices used for different ailments in dairy animals in peri-urban areas of Faisalabad (Pakistan). Int J Agric Biol 2009; 11(5): 535-41.

[169] Suthari S, Sreeramulu N, Omkar K, Raju V. The climbing plants of northern Telangana in India and their ethnomedicinal and economic uses. Indian J Plant Sci 2014; 3(1): 86-100.

[170] Ye H, Li C, Ye W, *et al.* Medicinal ferns of equisetaceae, angiopteridaceae, osmundaceae, lygodiaceae, *etc>*. Common Chinese Materia Medica 2021; 1: 47-68.
[http://dx.doi.org/10.1007/978-981-16-2062-1_4]

[171] Sureshkumar J, Silambarasan R, Bharati KA, Krupa J, Amalraj S, Ayyanar M. A review on ethnomedicinally important pteridophytes of India. J Ethnopharmacol 2018; 219: 269-87.
[http://dx.doi.org/10.1016/j.jep.2018.03.024] [PMID: 29578072]

[172] Morris JA. A molecular phylogeny of the Lythraceae and inference of the evolution of heterostyly. Kent State University 2007.

[173] Nagel JM, Griffin KL. Construction cost and invasive potential: comparing *Lythrum salicaria* (Lythraceae) with co-occurring native species along pond banks. Am J Bot 2001; 88(12): 2252-8.
[http://dx.doi.org/10.2307/3558387] [PMID: 21669658]

[174] Piwowarski JP, Granica S, Kiss AK. Lythrum salicaria L.—Underestimated medicinal plant from European traditional medicine. A review. J Ethnopharmacol 2015; 170: 226-50.
[http://dx.doi.org/10.1016/j.jep.2015.05.017] [PMID: 25985768]

[175] Sukenti K, Hakim L, Indriyani S, Purwanto Y, Matthews PJ. Ethnobotanical study on local cuisine of the Sasak tribe in Lombok Island, Indonesia. Journal of Ethnic Foods 2016; 3(3): 189-200.
[http://dx.doi.org/10.1016/j.jef.2016.08.002]

[176] Aziz MA, Adnan M, Khan AH, Sufyan M, Khan SN. Cross-cultural analysis of medicinal plants commonly used in ethnoveterinary practices at South Waziristan Agency and Bajaur Agency, Federally Administrated Tribal Areas (FATA), Pakistan. J Ethnopharmacol 2018; 210: 443-68.
[http://dx.doi.org/10.1016/j.jep.2017.09.007] [PMID: 28917974]

[177] Owino RA. Ethnobotany and mineral contents of indigenous vegetables of kisumu district in Kenya. University of Nairobi 1998.

[178] Hammond ME, Pokorný R, Okae-Anti D, Gyedu A, Obeng IO. The composition and diversity of natural regeneration of tree species in gaps under different intensities of forest disturbance. J For Res 2021; 32(5): 1843-53.
[http://dx.doi.org/10.1007/s11676-020-01269-6]

[179] Patel R, Riya K, Neha P. Floristic diversity of family malvaceae found in kankariya zoo campus. Ahmedabad, Gujarat, India 2018.

[180] Elkhalifa AEO, Alshammari E, Adnan M, *et al.* Okra (Abelmoschus esculentus) as a potential dietary medicine with nutraceutical importance for sustainable health applications. Molecules 2021; 26(3): 696.
[http://dx.doi.org/10.3390/molecules26030696] [PMID: 33525745]

[181] Das S, Nandi G, Ghosh L. Okra and its various applications in drug delivery, food technology, health care and pharmacological aspects-a review. Journal of Pharmaceutical Sciences and Research 2019; 11(6): 2139-47.

[182] Evans WC. Trease and Evans' pharmacognosy: Elsevier Health Sciences; 2009.

[183] Amusa MOO. A multidisciplinary study of South African Meliaceae. South Africa: University of Johannesburg 2021.

[184] Hossain MA, Al-Toubi WAS, Weli AM, Al-Riyami QA, Al-Sabahi JN. Identification and characterization of chemical compounds in different crude extracts from leaves of Omani neem. J Taibah Univ Sci 2013; 7(4): 181-8.
[http://dx.doi.org/10.1016/j.jtusci.2013.05.003]

[185] Rahmani A, Almatroudi A, Alrumaihi F, Khan A. Pharmacological and therapeutic potential of neem (Azadirachta indica). Pharmacogn Rev 2018; 12(24): 250-5.
[http://dx.doi.org/10.4103/phrev.phrev_8_18]

[186] Lawal A, Adekunle VAJ, Onokpise OU. Phytotherapy and polycyclic logging: implication on genetic multiplicity and diversity of african mahogany in tropical rainforest. Not Sci Biol 2019; 11(1): 21-5.
[http://dx.doi.org/10.15835/nsb11110373]

[187] McGaw LJ, Famuyide IM, Khunoana ET, Aremu AO. Ethnoveterinary botanical medicine in South

Africa: A review of research from the last decade (2009 to 2019). J Ethnopharmacol 2020; 257: 112864.
[http://dx.doi.org/10.1016/j.jep.2020.112864] [PMID: 32302713]

[188] Antonelli A, Smith R, Fry C, *et al.* State of the world's plants and fungi: Royal botanic gardens. Kew: Sfumato Foundation 2020.

[189] Ijaz F, Rahman IU, Iqbal Z, Alam J, Ali N, Khan SM. Ethno-ecology of the healing forests of Sarban Hills, Abbottabad, Pakistan: an economic and medicinal appraisal. Plant and Human Health, Volume 1: Ethnobotany and Physiology. 2018: 675-706.

[190] Khasim SM, Long C, Thammasiri K, Lutken H. Medicinal plants: biodiversity, sustainable utilization and conservation. Springer 2020.
[http://dx.doi.org/10.1007/978-981-15-1636-8]

[191] Lian L, Ortiz RDC, Jabbour F, Chen ZD, Wang W. Re-delimitation of *Tinospora* (Menispermaceae): Implications for character evolution and historical biogeography. Taxon 2019; 68(5): 905-17.
[http://dx.doi.org/10.1002/tax.12126]

[192] Galav P, Jain A, Katewa S. Ethnoveterinary medicines used by tribals of Tadgarh-Raoli wildlife sanctuary. Rajasthan, India 2013.

[193] Sharma R, Manhas RK, Magotra R. Ethnoveterinary remedies of diseases among milk yielding animals in Kathua, Jammu and Kashmir, India. J Ethnopharmacol 2012; 141(1): 265-72.
[http://dx.doi.org/10.1016/j.jep.2012.02.027] [PMID: 22366093]

[194] Upadhyay B, Singh KP, Kumar A. Ethno-veterinary uses and informants consensus factor of medicinal plants of Sariska region, Rajasthan, India. J Ethnopharmacol 2011; 133(1): 14-25.
[http://dx.doi.org/10.1016/j.jep.2010.08.054] [PMID: 20817085]

[195] Murray N.J., Keith D.A., Tizard R., *et al.* Threatened ecosystems of Myanmar. An IUCN Red List of ecosystems assessment. Version 1.0. 2020.

[196] Kokou K, Adjossou K, Kokutse AD. Considering sacred and riverside forests in criteria and indicators of forest management in low wood producing countries: The case of Togo. Ecol Indic 2008; 8(2): 158-69.
[http://dx.doi.org/10.1016/j.ecolind.2006.11.008]

[197] Hussain K, Shahazad A, Zia-ul-Hussnain S. An ethnobotanical survey of important wild medicinal plants of Hattar district Haripur, Pakistan. Ethnobotanical leaflets. 2008; 2008(1): 5.

[198] Hnini M, Taha K, Aurag J. Botany, associated microbiota, traditional medicinal uses, and phytochemistry of Vachellia tortilis subsp. raddiana (Savi): A systematic review. Journal of Agriculture and Food Research. 2023: 100566.

[199] Stark T, Mtui D, Balemba O. Ethnopharmacological survey of plants used in the traditional treatment of gastrointestinal pain, inflammation and diarrhea in Africa: future perspectives for integration into modern medicine. Animals (Basel) 2013; 3(1): 158-227.
[http://dx.doi.org/10.3390/ani3010158] [PMID: 26487315]

[200] Abba A, Dogara AM. Ethnomedicinal survey of plants used for management of inflammatory diseases in Ringim local government, Jigawa state, Nigeria. Ethnobot Res Appl 2021; 22: 1-27.
[http://dx.doi.org/10.32859/era.22.47.1-27]

[201] Kamatchi A, Parvathi AS. Quantitative analysis in traditional knowledge of wild medicinal plants used to treat livestock diseases by The Paliyar's Tribe of Sadhuragiri Hillstamil Nadu, India. Asian Journal of Pharmaceutical Research and Development 2020; 8(4): 44-57.

[202] Sindhu SS, Sharma R, Sindhu S, Sehrawat A. Soil fertility improvement by symbiotic rhizobia for sustainable agriculture. Soil fertility management for sustainable development. 2019: 101-66.
[http://dx.doi.org/10.1007/978-981-13-5904-0_7]

[203] Bahadur B, Reddy KJ, Rao M. Medicinal plants: an overview. Adv Med plants Univ Press Hyderabad.

2007.

[204] Wang HF, Xu X, Cheng XL, *et al*. Spatial patterns and determinants of Moraceae richness in China. J Plant Ecol 2022; 15(6): 1142-53.
[http://dx.doi.org/10.1093/jpe/rtac025]

[205] Wijaya IMS, Defiani MR. Diversity and distribution of figs (Ficus: Moraceae) in Gianyar District, Bali, Indonesia. Biodiversitas (Surak) 2020; 22(1).
[http://dx.doi.org/10.13057/biodiv/d220129]

[206] Williams EW, Gardner EM, Harris R III, Chaveerach A, Pereira JT, Zerega NJC. Out of Borneo: biogeography, phylogeny and divergence date estimates of *Artocarpus* (Moraceae). Ann Bot (Lond) 2017; 119(4): mcw249.
[http://dx.doi.org/10.1093/aob/mcw249] [PMID: 28073771]

[207] Koné WM, Kamanzi Atindehou K. Ethnobotanical inventory of medicinal plants used in traditional veterinary medicine in Northern Côte d'Ivoire (West Africa). S Afr J Bot 2008; 74(1): 76-84.
[http://dx.doi.org/10.1016/j.sajb.2007.08.015]

[208] Siddique Z, Shad N, Shah GM, *et al*. Exploration of ethnomedicinal plants and their practices in human and livestock healthcare in Haripur District, Khyber Pakhtunkhwa, Pakistan. J Ethnobiol Ethnomed 2021; 17(1): 55.
[http://dx.doi.org/10.1186/s13002-021-00480-x] [PMID: 34496911]

[209] Thorn JPR, Thornton TF, Helfgott A, Willis KJ. Indigenous uses of wild and tended plant biodiversity maintain ecosystem services in agricultural landscapes of the Terai Plains of Nepal. J Ethnobiol Ethnomed 2020; 16(1): 33.
[http://dx.doi.org/10.1186/s13002-020-00382-4] [PMID: 32513199]

[210] Ladio AH, Lozada M. Human ecology, ethnobotany and traditional practices in rural populations inhabiting the Monte region: Resilience and ecological knowledge. J Arid Environ 2009; 73(2): 222-7.
[http://dx.doi.org/10.1016/j.jaridenv.2008.02.006]

[211] Dirzo R. Seasonally dry tropical forests: ecology and conservation. Island Press 2011.
[http://dx.doi.org/10.5822/978-1-61091-021-7]

[212] Tripathi S, Singh S, Roy RK. Pollen morphology of Bougainvillea (Nyctaginaceae): A popular ornamental plant of tropical and sub-tropical gardens of the world. Rev Palaeobot Palynol 2017; 239: 31-46.
[http://dx.doi.org/10.1016/j.revpalbo.2016.12.006]

[213] Abd El-Ghani MM. Traditional medicinal plants of Nigeria: an overview. Agric Biol J N Am 2016; 7(5): 220-47.

[214] Wani ZA, Farooq A, Sarwar S, *et al*. Scientific appraisal and therapeutic properties of plants utilized for veterinary care in Poonch district of Jammu and Kashmir, India. Biology (Basel) 2022; 11(10): 1415.
[http://dx.doi.org/10.3390/biology11101415] [PMID: 36290318]

[215] Wilson KA, Wood CE Jr. The genera of Oleaceae in the southeastern United States. J Arnold Arbor 1959; 40(4): 369-84.
[http://dx.doi.org/10.5962/p.36686]

[216] Muthee JK, Gakuya DW, Mbaria JM, Kareru PG, Mulei CM, Njonge FK. Ethnobotanical study of anthelmintic and other medicinal plants traditionally used in Loitoktok district of Kenya. J Ethnopharmacol 2011; 135(1): 15-21.
[http://dx.doi.org/10.1016/j.jep.2011.02.005] [PMID: 21349318]

[217] González-Minero FJ, Bravo-Díaz L. The use of plants in skin-care products, cosmetics and fragrances: Past and present. Cosmetics 2018; 5(3): 50.
[http://dx.doi.org/10.3390/cosmetics5030050]

[218] Akerreta S, Calvo MI, Cavero RY. Ethnoveterinary knowledge in navarra (Iberian Peninsula). J

Ethnopharmacol 2010; 130(2): 369-78.
[http://dx.doi.org/10.1016/j.jep.2010.05.023] [PMID: 20573568]

[219] Khan M, Ahmad L, Rashid W. Ethnobotanical documentation of traditional knowledge about medicinal plants used by indigenous people in Talash valley of Dir lower. J Intercult Ethnopharmacol 2018; 7(1): 1.
[http://dx.doi.org/10.5455/jice.20171011075112]

[220] Stebbins GL. Flowering plants: evolution above the species level. Harvard University Press 1974.
[http://dx.doi.org/10.4159/harvard.9780674864856]

[221] Pattanayak S, Mandal TK, Bandyopadhyay S. A study on use of plants to cure enteritis and dysentery in three southern districts of West Bengal, India. J Medic Plants Studies 2015; 3(5): 277-83.

[222] Garden MP. Guide to the Medicinal Plant Garden. Indiana Medical History Museum 2010; 46222(317): 22.

[223] Borgohain P, Hazarika K. Plant diversity with especial reference to important medicinal plants in the Bhogdoi river of Jorhat, Assam, North East India.

[224] Ghorbanpour M, Hadian J, Nikabadi S, Varma A. Importance of medicinal and aromatic plants in human life. Medicinal plants and environmental challenges. 2017: 1-23.
[http://dx.doi.org/10.1007/978-3-319-68717-9_1]

[225] Ramawat K, Dass S, Mathur M. The chemical diversity of bioactive molecules and therapeutic potential of medicinal plants. Herbal drugs: ethnomedicine to modern medicine. 2009: 7-32.
[http://dx.doi.org/10.1007/978-3-540-79116-4_2]

[226] Zulak KG, Liscombe DK, Ashihara H, Facchini PJ. Alkaloids. Plant secondary metabolites: occurrence, structure and role in the human diet. 2006: 102-36.

[227] Ouvrard P, Transon J, Jacquemart AL. Flower-strip agri-environment schemes provide diverse and valuable summer flower resources for pollinating insects. Biodivers Conserv 2018; 27(9): 2193-216.
[http://dx.doi.org/10.1007/s10531-018-1531-0]

[228] Mawa S, Husain K, Jantan I. Ficus carica L.(Moraceae): phytochemistry, traditional uses and biological activities. Evidence-Based Complementary and Alternative Medicine. 2013; 2013.

[229] Setia A, Jangdey MS, Sharma M, *et al.* Holarrhena floribunda: A Potential Traditional Plant. Pharmaceutical and Biosciences Journal 2021; pp. 41-8.

[230] Sopeyin A, Ajayi G. Pharmacognostic study of Parquetina nigrescens (Afzel.) bullock (Periplocaceae). International Journal of Pharmacognosy and Phytochemical Research 2016; 8(2): 321-6.

[231] Pandit P. Inventory of ethno veterinary medicinal plants of Jhargram Division, West Bengal, India. Indian For 2010; 136(9): 1183.

[232] Sen S, Rengaian G. A review on the ecology, evolution and conservation of Piper (Piperaceae) in India: future directions and opportunities. Bot Rev 2021; 1-26.

[233] Tebbs M. Piperaceae Flowering plants· dicotyledons: magnoliid, hamamelid and caryophyllid families. Springer 1993; pp. 516-20.
[http://dx.doi.org/10.1007/978-3-662-02899-5_60]

[234] Titanji VP, Zofou D, Ngemenya MN. The antimalarial potential of medicinal plants used for the treatment of malaria in Cameroonian folk medicine. Afr J Tradit Complement Altern Med 2008; 5(3): 302-21.
[PMID: 20161952]

[235] Yuan C, Wang X, Wu J. Piper methysticum. Psychoactive Herbs in Veterinary Behavior Medicine 2008; p. 109.

[236] Salehi B, Zakaria ZA, Gyawali R, *et al.* Piper species: A comprehensive review on their phytochemistry, biological activities and applications. Molecules 2019; 24(7): 1364.

[http://dx.doi.org/10.3390/molecules24071364] [PMID: 30959974]

[237] Phondani PC, Maikhuri RK, Kala CP. Ethnoveterinary uses of medicinal plants among traditional herbal healers in Alaknanda catchment of Uttarakhand, India. Afr J Tradit Complement Altern Med 2010; 7(3): 195-206.
[http://dx.doi.org/10.4314/ajtcam.v7i3.54775] [PMID: 21461146]

[238] de Figueiredo R, Sazima M. Pollination biology of Piperaceae species in southeastern Brazil. Ann Bot (Lond) 2000; 85(4): 455-60.
[http://dx.doi.org/10.1006/anbo.1999.1087]

[239] Gibson DJ. Grasses and grassland ecology. Oxford University Press 2009.

[240] Lack A. Plant ecology and conservation: Garland Science; 2022.

[241] Fan J, Guo XY, Huang F, *et al.* Epiphytic colonization of*U stilaginoidea virens* on biotic and abiotic surfaces implies the widespread presence of primary inoculum for rice false smut disease. Plant Pathol 2014; 63(4): 937-45.
[http://dx.doi.org/10.1111/ppa.12167]

[242] Benjamin MAZ, Saikim FH, Ng SY, Rusdi NA. A comprehensive review of the ethnobotanical, phytochemical, and pharmacological properties of the genus Bambusa. Journal of Applied Pharmaceutical Science. 2023; 13(5): 001-22.

[243] Sarkar A. Ethnobotanical studies of sub-himalayan duars in West Bengal and assam with particular reference to the tribe mech. University of North Bengal 2011.

[244] Mphephu TS. Sustainable natural resource utilisation: a case study of ethnobotanically important plant taxa in the Thulamela Local Municipality, Limpopo Province. South Africa: University of Johannesburg 2017.

[245] Ribeiro C, Marinho C, Teixeira S. Uncovering the neglected floral secretory structures of Rhamnaceae and their functional and systematic significance. Plants 2021; 10(4): 736.
[http://dx.doi.org/10.3390/plants10040736] [PMID: 33918788]

[246] Donoghue MJ. A phylogenetic perspective on the distribution of plant diversity. Proc Natl Acad Sci USA 2008; 105(Suppl 1) (Suppl. 1): 11549-55.
[http://dx.doi.org/10.1073/pnas.0801962105] [PMID: 18695216]

[247] González-Stuart AE, Dhan Prakash DP, Charu Gupta CG. Phytochemistry of plants used in traditional medicine. 2014.

[248] Srinivas G, Babykutty S, Sathiadevan PP, Srinivas P. Molecular mechanism of emodin action: Transition from laxative ingredient to an antitumor agent. Med Res Rev 2007; 27(5): 591-608.
[http://dx.doi.org/10.1002/med.20095] [PMID: 17019678]

[249] Weerasuriya NM. Fungi Associated with Common Buckthorn (Rhamnus cathartica) in Southern Ontario. Canada: The University of Western Ontario 2017.

[250] Davis AP, Govaerts R, Bridson DM, Ruhsam M, Moat J, Brummitt NA. A global assessment of distribution, diversity, endemism, and taxonomic effort in the Rubiaceae1. Ann Mo Bot Gard 2009; 96(1): 68-78.
[http://dx.doi.org/10.3417/2006205]

[251] Feenna OP, Estella OU, Obodike EC. Phytochemical analysis and anti-diabetic activity of leaf extract of psydrax horizontalis schum and thonn (Rubiaceae). Pharmacogn J 2020; 12(1): 95-102.
[http://dx.doi.org/10.5530/pj.2020.12.15]

[252] Bertania S, Bourdyb G, Landaua I, Robinsonc JC, Esterred P, Deharo E. Evaluation of french guiana traditional antimalarial remedies. J Ethnopharmacol 2005; 98(1-2): 45-54.
[http://dx.doi.org/10.1016/j.jep.2004.12.020] [PMID: 15849870]

[253] Stefanello N, Spanevello RM, Passamonti S, *et al.* Coffee, caffeine, chlorogenic acid, and the purinergic system. Food Chem Toxicol 2019; 123: 298-313.

[http://dx.doi.org/10.1016/j.fct.2018.10.005] [PMID: 30291944]

[254] Tabuti JRS, Dhillion SS, Lye KA. Ethnoveterinary medicines for cattle (Bos indicus) in Bulamogi county, Uganda: plant species and mode of use. J Ethnopharmacol 2003; 88(2-3): 279-86.
[http://dx.doi.org/10.1016/S0378-8741(03)00265-4] [PMID: 12963156]

[255] Hilou A, Rappez F, Duez P. Ethnoveterinary management of cattle helminthiasis among the Fulani and the Mossi (Central Burkina Faso): plants used and modes of use. Int J Biol Chem Sci 2015; 8(5): 2207-21.
[http://dx.doi.org/10.4314/ijbcs.v8i5.24]

[256] Jinous Asgarpanah , Khoshkam R. Phytochemistry and pharmacological properties of Ruta graveolens L. J Med Plants Res 2012; 6(23): 3942-9.
[http://dx.doi.org/10.5897/JMPR12.040]

[257] Goldblatt P, Manning JC. Plant diversity of the Cape region of southern Africa. Ann Mo Bot Gard 2002; 89(2): 281-302.
[http://dx.doi.org/10.2307/3298566]

[258] Sonneman T. Lemon: A global history: Reaktion Books; 2013.

[259] Pollio A, De Natale A, Appetiti E, Aliotta G, Touwaide A. Continuity and change in the Mediterranean medical tradition: Ruta spp. (rutaceae) in Hippocratic medicine and present practices. J Ethnopharmacol 2008; 116(3): 469-82.
[http://dx.doi.org/10.1016/j.jep.2007.12.013] [PMID: 18276094]

[260] Maphosa V, Masika PJ. Ethnoveterinary uses of medicinal plants: A survey of plants used in the ethnoveterinary control of gastro-intestinal parasites of goats in the Eastern Cape Province, South Africa. Pharm Biol 2010; 48(6): 697-702.
[http://dx.doi.org/10.3109/13880200903260879] [PMID: 20645744]

[261] Chidambaram K, Alqahtani T, Alghazwani Y, Aldahish A, Annadurai S, Venkatesan K, et al. Medicinal plants of Solanum species: the promising sources of phyto-insecticidal compounds. Journal of Tropical Medicine. 2022; 2022.
[http://dx.doi.org/10.1155/2022/4952221]

[262] Alamgir A, Alamgir A. Medicinal, non-medicinal, biopesticides, color-and dye-yielding plants; secondary metabolites and drug principles; significance of medicinal plants; use of medicinal plants in the systems of traditional and complementary and alternative medicines (CAMs) Therapeutic Use of Medicinal Plants and Their Extracts. Pharmacognosy 2017; 1: pp. 61-104.
[http://dx.doi.org/10.1007/978-3-319-63862-1_3]

[263] Van Wyk B-E, Wink M. Medicinal plants of the world: Cabi; 2018.

[264] Pandey SK, Yadav SK, Singh VK. An overview on Capsicum annuum L. L. J Pharm Sci Technol 2012; 4(2): 821-8.

[265] Tewari D, Sah AN, Bawari S, Bussmann RW. Ethnobotanical investigations on plants used in folk medicine by native people of Kumaun Himalayan Region of India. Ethnobot Res Appl 2020; 20: 1-35.
[http://dx.doi.org/10.32859/era.20.16.1-35]

[266] Maggi F, Benelli G. Essential oils from aromatic and medicinal plants as effective weapons against mosquito vectors of public health importance. Mosquito-borne Diseases: Implications for Public Health. 2018: 69-129.
[http://dx.doi.org/10.1007/978-3-319-94075-5_6]

[267] Viljoen A, Mncwangi N, Vermaak I. Anti-inflammatory iridoids of botanical origin. Curr Med Chem 2012; 19(14): 2104-27.
[http://dx.doi.org/10.2174/092986712800229005] [PMID: 22414102]

[268] El-Din MIG, Fahmy NM, Wu F, *et al.* Comparative LC–LTQ–MS–MS analysis of the leaf extracts of Lantana camara and Lantana montevidensis growing in Egypt with insights into their antioxidant, anti-inflammatory, and cytotoxic activities. Plants 2022; 11(13): 1699.

[http://dx.doi.org/10.3390/plants11131699] [PMID: 35807651]

[269] Hassen A, Muche M, Muasya AM, Tsegay BA. Exploration of traditional plant-based medicines used for livestock ailments in northeastern Ethiopiaby. S Afr J Bot 2022; 146: 230-42.
[http://dx.doi.org/10.1016/j.sajb.2021.10.018]

[270] Venkatanarayana N, Basha G, Pokala N, Jayasree T, John SP, Nagesh C. Evaluation of anticonvulsant activity of ethanolic extract of Zingiber officinale in Swiss albino rats. J Chem Pharm Res 2013; 5(9): 60-4.

[271] Sahoo JP, Behera L, Praveena J, *et al.* The golden spice turmeric (Curcuma longa) and its feasible benefits in prospering human health—a review. Am J Plant Sci 2021; 12(3): 455-75.
[http://dx.doi.org/10.4236/ajps.2021.123030]

SUBJECT INDEX

A

Abdominal 46, 62
  disorders 62
  pain 46
Aesthetics, decorative 47
Africa, desert 93
Aging effects 54
Agricultural processes 4
Ailments 3, 9, 10, 22, 27, 34, 52, 72, 73, 76, 77, 87, 113, 115, 120, 121
  fluid-related 27
  gastrointestinal 22, 76
  menstrual 34, 113
  respiratory 10, 72, 115
Air purification 114
  podophyllum gusoot araceae 114
  woodii dil bel asclepiadaceae 114
Anemia 54, 114
Anethum graveolens 31
Animal 15, 56
  disease treatment, domestic 56
  husbandary 15
Animal health 55, 59, 60, 63, 65, 66, 69, 73, 81, 89, 90, 100, 101, 105
  management 105
  treatment 63, 65, 66, 73, 81, 89, 90
Animal illnesses 71, 93
  domestic 71
Annual herbs 27, 32, 63
Anthelmintic cure 91
Anti-inflammatory 22, 23, 34, 76, 109
  agent 23
  attributes 22, 34
  effects 76, 109
Anti-microbial properties 54
Anti-oxidant properties 44
Anti-termite activity 114
Antibacterial properties 71
Antioxidant(s) 3, 5, 59, 71, 72, 73, 105, 110, 114
  effects 59

properties 110
Antipyretic effects 87
Aromatherapy 72
Aromatic smell 33
Arthritis 27, 76, 77, 84, 110, 111, 115, 117, 120
  pain 111, 117
Ayurveda and unani systems of medical tradition 2
Ayurvedic 2, 8, 9
  practitioners 8
  products 9
  treatment 8

B

Bermuda grass 102, 103
Bignoniaceae plants 51
Biodiversity 8, 12, 21, 93, 95, 98, 100, 107, 112, 118
  boosting 93
  hotspot 12, 21, 112
Blood 3, 5, 30, 36, 106, 113, 117
  coagulation 3, 5
  dysentery 30
  purification 36, 113
  purifier 106, 117
Body inflammation 97
Bone fracture 58, 74, 114, 115, 120, 121
Boraginaceae plants 55
Botanical diversity 100, 101, 109, 110, 112
Bronchitis 43, 72, 114
Burning 41, 108
  seeds 108
  sensation 41
Burns, small 49
Butea monosperma 70

C

Cardiac glycosides, dangerous 38
Cardiovascular troubles 37

www.ingramcontent.com/pod-product-compliance
Lightning Source LLC
Chambersburg PA
CBHW041444210326
41599CB00004B/126